EYEWITNESS TRAVEL

TOP 10

GREEK ISLANDS

W9-DDG-069

CAROLE FRENCH

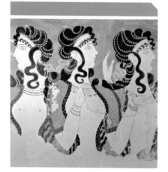

Top 10 Greek Islands Highlights

The Top 10 of Everything

DK | Penguin
        Random
        House

CONTENTS

Greek Islands Area by Area

Streetsmart

Within each Top 10 list in this book, no hierarchy of quality or popularity is implied. All 10 are, in the editor's opinion, of roughly equal merit.

Front cover and spine *Blue-domed church perched on the hillside in Oía, Santoríni*
Back cover *Stunning indigo waters and limestone cliffs at Navágio Beach, Zákynthos*
Title page *Windmills dotting the landscape of Sérifos in the Cyclades*

Welcome to
The Greek Islands

With endless days of sunshine, dazzlingly beautiful beaches, magnificent countryside and legendary historic sites, the Greek Islands have it all. These stunning islands make for a Mediterranean holiday like no other, whether you're seeking a beach holiday, hiking destination or cultural break. With Eyewitness Top 10 Greek Islands, they are yours to explore.

Elegant **Corfu Old Town** and atmospheric **Argostóli**, the capital of **Kefalloniá**, are just two of the must-see sights of the Ionians. To the south are the unspoiled Argo-Saronic Islands, home to the magnificent **Temple of Aphaia** on **Égina**. The Cyclades and the Sporádes include the blue and white splendour of **Santoríni**, whimsical **Mýkonos** – with the sacred city of **Delos** close to its shores – and quintessential **Skiáthos**, a magnet for film-makers.

Admire the artistic legacy of the Knights Hospitallers in **Rhodes Old Town** and visit the **Monastery of St John** where the saint wrote the Book of Apocalypse. See Macedonian influences at the **Néa Moní** on the Northeast Aegean island of **Híos**, explore ancient **Pythagóreio and Heraion** on neighbouring **Sámos**, and don't leave without visiting **Crete**, where you can walk in the footsteps of ancient civilizations at the **Palace of Knossos**. Along the way you'll find superb resorts with watersports, restaurants and nightlife.

Whether you're coming for a weekend or a week, our Top 10 guide brings together the best of everything the islands have to offer. You will find tips throughout, from seeking out what's free to finding the best restaurants, along with nine easy-to-follow itineraries designed to help you visit a clutch of sights in a short space of time. Add inspiring photography and detailed maps, and you've got the essential pocket-sized travel companion. **Enjoy the book, and enjoy the Greek Islands**.

Clockwise from top: **Oía, Santoríni; Palace of the Grand Masters, Rhodes; Pórto Katsíki beach, Lefkáda; Irákleio harbour, Crete;** *pithoi* at Delos; dovecotes, Tínos; fresco at the Palace of Knossos, Crete

Exploring the Greek Islands

From ancient palaces to rugged coastlines, unspoiled villages and countryside blanketed by vineyards, the Greek Islands are rich in beauty and culture. A two-day visit to Corfu with Paxí gives a taste of the Ionians, while highlights of the other island groups can be enjoyed with a longer stay.

The Achílleion Palace on Corfu, where statues of Greek gods take centre stage, is a splendid place to visit.

Key
— Two-day itinerary
— Seven-day itinerary

Two Days in Corfu and Paxí

Day ❶
MORNING
Start on **Corfu** (see p76) among the cobbled streets of **Corfu Old Town** (see pp12–13). See Plateía Spianáda before visiting the Dimarchío. Enjoy lunch and people-watching at a trendy café in the Listón.

AFTERNOON
Marvel at the red-topped campanile of the Church of Ágios Spyrídon. On the seafront, visit the Palace of St Michael and St George to see its fabulous Asiatic art. Outside, check out the view of the Palaió Froúrio.

Day ❷
MORNING
Take the fast hydrofoil to Gäios on the island of **Paxí** (see p75). A taxi tour here will give you a real taste of Ionian rural life. Stop at Lákka for an early fish lunch before returning to Corfu for the afternoon.

AFTERNOON
Join a tour to the **Achílleion Palace** (see p82) and revel in its gardens canopied by palm trees. Then go to **Taverna Sebastian** (see p83) for dinner.

Seven Days Around Rhodes, Sámos, Égina, Évvia, Crete and Mýkonos

Day ❶
In **Rhodes Old Town** (see pp14–15), head straight for the must-see sights: the Palace of the Grand Masters and the Street of the Knights. Have lunch at **Alexis 4 Seasons** (see p117) and drive to the ancient **Líndos acropolis** (see p116). End the day with dinner at **Kalypso Roof Garden** (see p117).

Mýkonos is identified by its many pretty windmills.

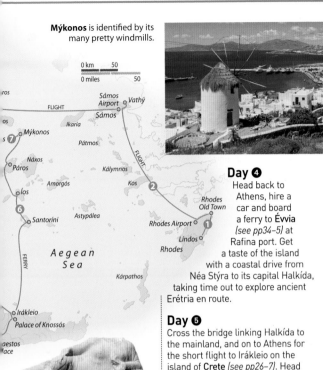

Day ❹
Head back to Athens, hire a car and board a ferry to **Évvia** *(see pp34–5)* at Rafina port. Get a taste of the island with a coastal drive from Néa Stýra to its capital Halkída, taking time out to explore ancient Erétria en route.

Day ❺
Cross the bridge linking Halkída to the mainland, and on to Athens for the short flight to Irákleio on the island of **Crete** *(see pp26–7)*. Head straight to the **Palace of Knossos** *(see pp30–31)* and look out for the Priest-King fresco as you explore the labyrinthine site. Next stop off at Phaestos Palace before heading back to Irákleio and feasting in **Erganos Tavern** *(see p108)*.

Day ❻
Catch the fast ferry from Irákleio to **Mýkonos** *(see p91)*, which calls in at **Santoríni** *(see p91)*, **Íos** *(see p88)* and **Páros** *(see p90)* along the way. On arrival enjoy a late lunch at **Raya Restaurant** *(see p95)* before heading inland to see the famous windmills of Mýkonos dotting the landscape.

At the Sanctuary of Heraion, statues mark the Sacred Way to the temple.

Day ❷
Board the short flight from Rhodes to **Sámos** *(see p125)* and head to **Pythagóreio** and the **Sanctuary of Heraion** *(see pp24–5)*. See the sites' treasures at the museum in Vathý.

Day ❸
Fly from Sámos to Athens and make your way to Piraeus port to take a ferry to **Égina** *(see p145)*. Spend the day exploring the charming island, stopping at the **Temple of Aphaia** *(see pp32–3)*, the **Temple of Apollo** *(see p154)* and the fishing village of **Pérdika** *(see p154)*. Finally, dine on fresh fish next to the harbour at **Nontas** *(see p155)*.

Day ❼
Board a boat from Mýkonos Town to **Delos** *(see pp18–19)*. Spend a leisurely day enthralled by the remains left by the ancient inhabitants in the **Maritime Quarter**, the **Sanctuary of Apollo** and the **Theatre Quarter**.

These itineraries focus on just a few of the Greek Islands and their highlights

Top 10 Greek Islands Highlights

Ladies in Blue fresco at the Palace of Knossos, Crete

TOP 10 Greek Islands Highlights

The Greek Islands comprise over 6,000 islands and islets, spread across several archipelagos. Some host holiday resorts, while others consist of rural communities. In others, ancient temples sit amid cosmopolitan towns. Add beaches, pine forests, olive groves, spectacular coves and bays and the result is mesmerizing.

1 Corfu Old Town

The arcaded terraces of the Listón, the ancient fortresses and museums, and Plateía Spianáda, with its Venetian architecture, all combine to give Corfu Old Town its charm *(see pp12–13)*.

2 Rhodes Old Town

This city was occupied by the Knights Hospitallers (1309–1522), who left such magnificent treasures as the Palace of the Grand Masters and the City Walls *(see pp14–15)*.

3 Monastery of St John, Pátmos

Dedicated to St John, who reputedly wrote the *Book of Revelation* nearby, this 11th-century monastery is a UNESCO site *(see pp16–17)*.

4 Delos

According to ancient Greek mythology, this tiny uninhabited island was the birthplace of Apollo and Artemis. It also has remains of civilizations dating from the 3rd century BC *(see pp18–19)*.

⑥ Pythagóreio and Heraion, Sámos

The remains of Pythagóreio, an ancient Greek and Roman fortified port, and Heraion, a Neolithic temple, have put Sámos on the heritage map *(see pp24–5)*.

⑤ Néa Moní, Híos

Containing one of Greece's finest collections of mosaics, this 11th-century monastery was built by Emperor Constantine IX Monomachos *(see pp20–21)*.

⑦ Crete

This popular Mediterranean holiday island is best known for its varied scenery and its Minoan archaeological sites, like the palaces of Knossos and Phaestos *(see pp26–9)*.

Lésvos

os

Híos

Néa Moní ⑤

Ándros

Pythagóreio and Heraion ⑥

os ④ Mýkonos

Cyclades

Náxos

③ Monastery of St John

Kos

Dodecanese

Astypálea

Rhodes Old Town ②

Santorini

Rhodes

Aegean Sea

0 km   100
0 miles   100

mno

⑦ ⑧ Palace of Knossos

ete

⑧ Palace of Knossos, Crete

The remains of the enormous Minoan Palace of Knossos give an incomparable insight into this ancient civilization *(see pp30–31)*.

⑨ Temple of Aphaia, Égina

A well-preserved Doric temple dedicated to Aphaia, the ancient Greek goddess of fertility, this structure dates from around 480 BC. It stands on a hilltop covered with pine trees on the island of Égina *(see pp32–3)*.

Évvia ⑩

This long, narrow and largely mountainous island has been ruled in turn by the Macedonians, Romans, Venetians and Ottoman Turks. Their influence gives Évvia its inimitable character and distinctive architecture *(see pp34–5)*.

TOP 10 ⭐ Corfu Old Town

With its cobbled plazas and tiny alleyways dating back to ancient times, Corfu Old Town continues to retain its old-world charm. Palaces, museums, fortresses, gourmet restaurants, traditional tavernas, cultural venues and a lively harbour combine to give the town its inimitable character. It has beautiful arcades reminiscent of the finest in Paris, and elegant Venetian mansions, which line the town's main thoroughfare, the Kapodistriou. Add Greek, Italian and British influences and you have an eclectic architectural anthology.

1 Listón
The Listón (above), built by the French in the early 1800s, was inspired by the grandiose buildings along the Rue de Rivoli in Paris. Its arcaded terraces were once used solely by Corfu's aristocrats, but are now full of stylish cafés where locals and visitors chill out.

2 Palace of St Michael and St George
Dating from the British period, this imposing Georgian-style palace has been the home of the Greek royal family, the British High Commission and the island's treasury. It is now home to the Museum of Asian Art.

3 Maitland Rotunda
Built to commemorate the life of Sir Thomas Maitland, the first Lord High Commissioner of the Ionian Islands during the British administration, this 19th-century monument (below) dominates the southern end of Plateía Spianáda.

NEED TO KNOW

MAP B5

Palace of St Michael and St George: Museum of Asian Art, Plateía Spianáda; 26610 30443; open Apr–Oct: 8am–8pm daily, Nov–Mar: 8am–4pm Tue–Sun; adm €6, concessions €3

Church of Ágios Spyrídon: Agíou Spyrídonos; open 6:30am–8pm daily (tourist visits are strongly discouraged during church services)

Néo Froúrio: Plateía Solomóu; open 8am–7:30pm daily (Nov–Mar: to 3pm)

Antivouniótissa Museum: Prosfórou 30; 26610 38313; open 8:30am–3pm Tue–Sun; adm €2, concessions €1

Palaió Froúrio: Palaió Froúrio Islet; 26610 48120; open 8am–8pm daily (Nov–Mar: to 3pm; adm €3, concessions €2

■ Enjoy a coffee at one of the chic cafés at the Listón.

4 Church of Agios Spyrídon

Named after Corfu's patron saint, St Spyrídon, whose remains lie here in a silver coffin, this 16th-century church has a distinct red-topped campanile with bells that ring at regular intervals.

7 Antivouniótissa Museum

With its Byzantine and post-Byzantine icons and ecclesiastical artifacts, this museum is in the Church of Panagía (Our Lady) Antivouniótissa, one of the city's oldest religious buildings.

5 Néo Froúrio

The Venetians built this mighty fort (below) in the 1500s. Despite its name, the New Fortress was completed only a few years after the Old Fortress. With a maze of medieval walkways, the former garrison is fun to explore.

THE MÁNOS COLLECTION

One of the key exhibits at the Museum of Asian Art in Corfu Old Town is the Mános Collection. Corfiot diplomat Grigórios Mános (1850–1929) had a passion for Japanese, Chinese and Korean art, furnishings, ceramics and weapons. He donated his fabulous private collection to the government to establish a museum in Corfu.

8 Plateía Spianáda

Venetian and French architecture give this huge bustling square an elegant feel. The maze of alleyways off the square, Campiello, is one of the town's oldest parts.

9 Dimarchío

The Dimarchío, or Town Hall, is a classic Venetian-style building that once served as the San Giacomo theatre, a favourite with the nobility.

Corfu Old Town

10 Palaió Froúrio

The Old Fortress was built in the 16th century on a tiny islet to protect the city from invaders. Its restored interior has become a popular venue for cultural events. At its base stands St George's Church.

6 The Old Port

While cruise ships head for the New Port these days, in ancient times the Old Port (right) would have been a hive of nautical activity. It is located between the fortresses and is in one of the most picturesque parts of the Old Town.

TOP 10 ⭐ Rhodes Old Town

The citadel at the heart of Rhodes Town, the capital of Rhodes island, is a "living museum" showcasing the ancient and the medieval areas of the city. The Knights Hospitallers occupied the city from 1309 and transformed it into a formidable stronghold, but in 1522 Suleiman the Magnificent, Sultan of the Ottoman Empire, conquered the Knights. The UNESCO World Heritage Site walled city boasts outstanding buildings from both periods, including the Palace of the Grand Masters, Street of the Knights, mosques and hammams.

Palace of the Grand Masters

Located in the citadel in the Collachium area, the current building **(right)** is a replica of the original Knights' palace that was destroyed in 1856. It is now a museum *(see p43)*.

NEED TO KNOW

MAP V4

Palace of the Grand Masters: Ippotón; 22413 65270; open Apr–Oct: 8am–8pm daily; Nov–Mar: 8am–3pm Tue–Sun; adm €6, concessions €3

Archaeological Museum, Rhodes: Hospital of the Knights, Plateía M. Alexandrou; 22413 65257; open Apr–Oct: 8am–8pm daily; Nov–Mar: 8am–3pm Tue–Sun; adm €8, concessions €4

Municipal Museum of Modern Greek Art: Symi Square; 22410 36646; open 8am–9pm Mon–Fri (winter to 1pm); adm €3, concessions €1

■ A special combined ticket is available that allows entry to some of the churches and museums here.

■ Head for the Plateía Ippokrátous, Plateía Martýron Evraíon and Sókra tou Sokrátous for the best choice of places to eat and drink.

② Street of the Knights

A cobbled street crossed by arched bridges and lined with the Inns of the Tongues, this site dates from the 14th century and is one of the world's finest examples of Gothic urbanism.

③ Hóra

The citadel is divided into two main areas, the Hóra (Bourg) and the Collachium, home to some of the busiest streets in the city. Cafés vie for attention with markets full of stalls.

Archaeological Museum, Rhodes ④

Known for its amphorae collection and artifacts **(right)**, this museum is housed in the Hospital of the Knights. Its infirmary is the main exhibition hall.

7 Marine Gate

Dominating the Plateía Ippokrátous, Pýli Agías Aikaterínis, also known as Marine or Sea Gate **(left)**, is a mighty bastion of two towers and is the most magnificent of all the gates leading to the inner Old Town. The gate was restored in 1951.

INNS OF THE TONGUES

Nationality was the basis on which the Knights Hospitallers, Order of St John of Jerusalem, were divided into groups known as "Tongues". Their meeting places or inns, with their coat of arms, lined the Street of the Knights. The Inn of Provence and Chapelle Française are on the street's north side, while on the south is Spain.

Rhodes Old Town

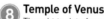

10 Jewish Quarter

This area dates from the 1st century AD. The Plateía Martýron Evraíon has a very moving monument to the Rhodian Jews who died in World War II concentration camps.

8 Temple of Venus

Thought to date from the 3rd century BC, this temple was identified as dedicated to Venus due to the nature of the votive offerings found by archaeologists. The remaining columns and fragments demonstrate that Rhodes was a town of major importance during the Hellenistic period as well as during medieval times.

5 Municipal Museum of Modern Greek Art

This impressive collection features a permanent exhibition of important contemporary Greek art as well as several special themed exhibitions throughout the year.

6 Medieval City Walls

The Knights Hospitallers built these walls on Byzantine fortifications to completely encircle and protect their city. The 4-km- (6-mile-) long walls contain a labyrinth of tiny alleyways.

9 Mosque of Suleiman the Magnificent

Erected as a triumphal mosque by Suleiman in 1522, the present structure **(below)** was built in 1808 using materials from the original.

TOP 10 ★ Monastery of St John, Pátmos

The island of Pátmos is reputedly where St John wrote the *Book of Revelation* in AD 95, the last book of the New Testament. John is said to have lived the life of a hermit in a cave in ancient Hóra, where he received apocalyptic visions from Jesus Christ that compelled him to write the work. The Cave of the Apocalypse lies close to the monastic complex of the Monastery of St John, built in 1088 in the saint's honour. Hóra, the cave and the monastery are a UNESCO World Heritage Site.

NEED TO KNOW

MAP F4

Monastery of St John: Hóra; 22470 31234; open 8am–1:30pm daily and 4–6pm Tue, Sat & Sun

Treasury: 22470 20800; open 8am–1:30pm daily and 4–6pm Tue, Sat & Sun; adm €2

■ Aim to arrive in the early morning or late afternoon to avoid the tour crowds and the heat of the midday sun.

■ Hóra has a few small cafés and tavernas for refreshments but be sure to take bottled water with you.

1 Hóra
This ancient settlement has over 40 monasteries and chapels and is one of a few places where early Christian ceremonies are still practised. Note the town's Byzantine architecture.

3 Icon of St John
The most sacred treasure here is the Icon of St John, which is housed in the inner narthex of the monastery. Dating from the 12th century, it shows the saint holding his work.

2 The Chapel of Christodoúlou
Ioánnis Christódoulos, the Blessed, was the monk who had the Monastery of St John built. This chapel **(below)** is dedicated to him and houses his remains in a marble sarcophagus.

7 Sanctuary of Dionysos

Dedicated to Dionysos, god of wine and ecstasy, and Zeus's son, this sanctuary **(left)** was used for ritual worship and is known for its 2,300-year-old phallic monuments.

5 Hall of the Poseidoniasts

The cultic hall housed meeting rooms belonging to Beirut merchants who worshipped Baal (known as Poseidon) here.

Delos

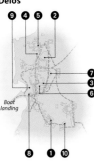

8 Agorá of the Competaliasts

The remains of shops and potholes for awnings can still be seen in the Hellenistic-era market-place, occupying an area near the Sacred Harbour. The stone-paved ground is heavily worn.

9 Maritime Quarter

Revolving around the Sacred Harbour, the Maritime Quarter was the main residential area of Delos. Among the ruins are floor mosaics of the mansions built by wealthy merchants. The House of Masks has a fine mosaic of Dionysus riding a panther.

6 Temple of the Delians

The classic Doric-style Temple of the Delians is one of the many excavated, along with the Roman-era Temple of Isis and the 5th-century Temple of Hera.

PEOPLE OF DELOS

Delos is uninhabited, but this was not always so. In 426 BC, the Athenians decided to "cleanse" the island and thousands of inhabitants were exiled. It was declared that no one would ever be born, die or be buried in this holy place. Graves were opened and the remains removed. As a result, the island was abandoned.

10 House of Dolphins

With its floor mosaic of dolphins at play **(above)**, this house to the amphi-theatre's north is very well preserved. Dating from the 2nd century BC, it gives an insight into the fashion of the day.

NEED TO KNOW

MAP P6

Delos Archaeological Museum: Delos; 22890 22259; open 8am–8pm daily; adm €5, concessions €3

■ There is hardly any shade, so wear sunscreen and a hat to guard against the sun. Also, take bottled water to avoid dehydration.

■ The Sanctuary of Apollo has a small restaurant that serves a range of refreshments.

TOP 10 ⭐ Néa Moní, Híos

Considered one of the finest examples of architecture from the Macedonian Renaissance, this UNESCO World Heritage Site is famous for its golden mosaics. Located just west of Híos Town, the 11th-century Néa Moní, meaning New Monastery, was built by the Byzantine Emperor Constantine IX Monomachos and Empress Zoe (r.1042–1050). According to legend, three monks found a miraculous icon of the Virgin Mary here. It became an influential and wealthy monastery but its decline began when the Ottomans plundered Híos in 1822.

1 Churches
The monastery has two small churches. One is dedicated to the Holy Cross **(above)**. The other, Panteleimon, is named after the saint who, according to legend, became a martyr during the Roman persecution of Christians.

2 The Narthex
The esonarthex (inner narthex) and the exonarthex (outer narthex) entrances feature some of the monastery's most prized mosaics. There is also a marble inlaid floor, which was a sign of wealth in Byzantine times.

3 Trapeza
The dining hall, or refectory, is where the monks would have met for meals. A once-massive structure, it was badly damaged in 1822 by the Ottomans and again by the 1881 earthquake. Its remains can be visited.

4 The Belfry
The original belfry was built in the early 16th century. It collapsed during the 1881 earthquake and was rebuilt in 1900 **(below)**.

NEED TO KNOW

MAP K5 ▪ Híos

Monastery: 9am–1pm and 4:30pm–dusk daily

Museum: 9am–1pm Tue–Sun; adm €2 (free on Sunday)

▪ Visitors are advised to dress conservatively. Shorts must be avoided, and a head scarf must be worn by women (these are available to borrow on site).

▪ The monastery is surrounded by small villages that have tavernas and coffee shops for refreshment.

7 The Catholicon
The main part of the church between the esonarthex and exonarthex **(left)** is known as the Catholicon, or Katholikón, and is lavishly decorated with marble and mosaics. Dedicated to the Virgin Mary, it is a rare early example of insular octagonal architecture and is dominated by its dome.

MACEDONIAN RENAISSANCE ART

The Macedonian Renaissance, a period of Byzantine art, began in the late 9th century and continued till the mid-11th century. During this time, cross-in-square and octagonal churches with fabulous frescoes, icons and mosaics were commissioned. The use of marble and ivory, often on gold and black backgrounds, became fashionable. The finest examples from the period are Ósios Loukás Moní and Dafní Moní on the Greek mainland, which share UNESCO status with Néa Moní.

9 Skull Cabinet
This cabinet contains skulls of the islanders massacred by the Ottomans in 1822, including 600 monks from Néa Moní **(below)**.

5 Mosaics
Worked in marble on a gold background, these exquisite mosaics depict biblical scenes and figures. The works include the famous *Anástasis* (Resurrection), showing Christ's rescue of Adam and Eve from Hell **(above)**. The collection is one of only three left in Greece of the mid-Byzantine period.

6 Cistern
A key part of the monastery's infrastructure, the cistern, or *kinsterna*, is a well-preserved underground complex of marble columns, arches and vaults, which was designed to collect and hold rain water. This water was then supplied to the monks.

8 St Luke's Chapel
This small chapel is dedicated to St Luke, the early Christian author of the *Gospel of Luke*, and is near the monks' cemetery, just outside the monastery's boundary wall. The chapel's architecture is characteristic of the ecclesiastical style dating from the 11th-century.

10 Monks' Cells
Venetian-period cells, known as Keliá, are where the monks would have slept. Though most are in a ruined state, one has been renovated and houses a small museum.

🔟 ⭐ Pythagóreio and Heraion, Sámos

Pythagóreio, named after the ancient philosopher Pythagóras, and the Sanctuary of Heraion are two treasures of Sámos. Pythagóreio, now a holiday resort, was an important port in antiquity, its defences acting as a stronghold against invasion. Remains of the town's citadel, Roman baths, the harbour and an aqueduct can be seen. Nearby, Heraion is a sanctuary where Hera, the Greek goddess of fertility, was cult worshipped. Its architecture is among the finest from the period.

1 Sanctuary of Heraion

Within the grounds of this large sanctuary, dedicated to Hera, the goddess of women and family, stand the archaeological remains of the Temple of Hera and an early Christian basilica. There are several statues **(right)** lining the Sacred Way path that leads to the temple.

2 Ancient Fortification

The city walls of Pythagóreio, known as Polykrates Wall, date from the 6th century BC and surround the ancient town and its harbour. They run from the Lykoúrgos Logothétis tower to the Tunnel of Eupalinos, a total circumference of more than 6 km (4 miles) with 12 gates.

3 Temple of Hera

A single column is all that remains of the temple, but signs show that it was once among the largest in antiquity. The goddess Hera is said to have been born, raised and worshipped here.

4 Ancient Harbour

Pythagóreio's harbour is considered the oldest man-made maritime installation in the Mediterranean. Dating from the 6th century BC, it is now silted up and can be crossed by a causeway.

5 Ancient Port

Sámos was a major naval and mercantile power in the 6th century BC. Its port **(left)** was once bustling with ships and merchants, through which produce, materials and grain were traded. It brought great wealth to the island.

8 Tunnel of Eupalinos

Constructed in the 6th century by ruler of Sámos, Polykrates, to provide his people with water, this extravagant aqueduct **(left)** is an engineering marvel. It is particularly notable as one of the first tunnels in history to be excavated from both ends.

PYTHAGÓRAS

Pythagóras was a 5th-century-BC Greek scientist, philosopher, and mathematician, who, legend says, was the son of Apollo. He is best known for his theorem about the geometry of triangles. He is also credited with Pythagórean tuning, a musical system where instruments are tuned to intervals on a 3:2 ratio, and founded a religious movement, Pythagóreanism.

6 Statue of Pythagóras

Standing on the harbour at Pythagóreio, this 3-m- (9-ft-) tall statue celebrates the life of Pythagóras the Samian, who was born on the island in 570 BC. The town is named after him.

9 Greek and Roman Remains

A wealth of Roman remains lie around Pythagóreio, including an amphitheatre and an acropolis located in the castle grounds.

10 Castle of Lykoúrgos Logothétis

This castle **(above)** and tower was built by a local hero, Lykoúrgos Logothétis, three years after his heroic part in the War of Independence of 1821. It is believed to stand on the site of the ancient acropolis.

7 Archaeological Museum of Sámos

Many Pythagóreio and Heraion treasures, including Koúros, a 5-m- (16-ft-) tall marble statue dedicated to the god Apollo and dating to 580 BC, are displayed in this wonderfully engaging museum in Vathý.

NEED TO KNOW

MAP F4

Archaeological Museum of Sámos: Plateia Dimarchiou, Vathý; 22730 27469; open 8am–3pm Tue–Sun; adm €4, concessions €2

■ There is little shade at the Sanctuary of Heraion and some of the other ancient remains around Pythagóreio, so wear a hat and carry water if visiting when it is hot.

■ When visiting the ancient site and the nearby sanctuary, visitors can stop for refreshments at Pythagóreio's many tavernas and restaurants.

★ Crete

The largest Greek island, Crete is one of Greece's most charming holiday destinations. The island is an intoxicating mixture of cosmopolitan cities such as Irákleio and Chaniá, sprawling resorts and dramatic scenery that encompasses mountains, gorges, rugged coastlines and a countryside blanketed by olive and citrus groves. Crete's temperate climate and sandy beaches draw holidaymakers throughout the summer, while hiking in the spectacular ravine, the Samariá Gorge, and exploring Crete's magnificent archaeological sites, including the world renowned Minoan palaces of Knossos *(see pp30–31)* and Phaestos, are popular attractions year-round.

1 Ágios Nikólaos

Hugging the shores of Mirabéllou Bay and the deep lake of Voulisméni, the picturesque harbour of Ágios Nikólaos **(above)** is one of Crete's most delightful holiday spots. It stands on the site of the ancient city of Lato *(see p104)*.

2 Vaï

This long stretch of beach is best known for the hundreds of Cretan Date Palm trees that line its route **(below)**. Dating from at least Classical times, it is Europe's largest natural palm forest *(see p103)*.

3 Palace of Mália

The remains of the Palace of Mália *(see p102)*, once a major city, are distinguished by Minoan architectural elements that include a central courtyard with a sacrificial altar, raised processional ways and a labyrinth of underground crypts. Some of these rooms still contain numerous *pithoi* **(left)**.

4 Amári Valley and Agía Triáda

Dotted with traditional villages, the Amári Valley has viewpoints to enjoy the panoramic scenery. To its east is Mount Idi, while the Minoan palace Agía Triáda is a short drive away to the south *(see p101)*.

6 Chaniá

Postcard-pretty Chaniá **(left)**, once Crete's capital, hugs the shore towards the western end of the island. The majestic Lefká Óri mountain range provides a dramatic backdrop to its scenic harbour and cobbled Splántzia Quarter *(see p102)*.

7 Phaestos Palace

The central courtyard and grand staircase are among the highlights of this Minoan palace, plus the remains of an earlier palace *(see p101)*. The Phaestos Disc was discovered here.

8 Réthymno

Once a thriving Minoan and later Venetian city, Réthymno *(see p102)* developed to become a fashionable haunt of artists, writers and scholars. Despite tourism, the city has retained much of its charm .

5 Irákleio

Venetian mansions, a formidable fortress, elegant restaurants and trendy boutiques, Crete's capital is a vibrant place *(see p102)*. The Irákleio Archaeological Museum *(see p28)* houses the world's finest collection of Minoan art.

9 Górtys

The remains of a theatre with rare inscriptions, a basilica and a citadel complete with an agora (marketplace) can be seen at the historic site of Górtys *(see p101)*.

THE PHAESTOS DISC

The Phaestos Disc was discovered in Phaestos Palace in 1903. Dating from the 17th century BC, this clay disk has the finest example of Minoan script ever found. The inscription uses 45 signs stamped into the clay by a seal, which are said to be religious text. It's now on display in the Irákleio Archaeological Museum.

10 Samariá Gorge

With its sheer rock faces and challenging terrain, the Samariá Gorge **(below)** is the longest ravine in Europe *(see p103)*. Chapels and villages occupied before the area became a national park can be seen here.

NEED TO KNOW

Palace of Mália: **MAP E6**; Mália; 28970 31597; open 8am–5pm daily; adm €6, concessions €3

Agía Triáda: **MAP E6**; near Phaestos; 28920 91564; open 9am–4pm daily; adm €4, concessions €2

Irákleio Archaeological Museum: **MAP E6**; Xanthoudidou 2, Irákleio; 28102 79000; open Apr–Oct: 8am–8pm daily, Nov–Mar: 11am–5pm Mon, 8am–3pm Tue–Sun; adm €10, concessions €5

Phaestos Palace: **MAP E6**; Phaestos, Mires, near Irákleio; 28920 42315; open 8am–8pm daily; adm €8, concessions €4

Górtys: **MAP E6**; Agioi Déka; 28920 31144; open 8am–8pm daily; adm €6, concessions €3

Samariá Gorge: **MAP D6**; 28210 45570; open May–Oct: 6am–4pm daily (subject to weather); adm €5

■ Combined tickets are available for Phaestos Palace and the Royal Villa at Agía Triáda, or the Palace of Knossos and the Irákleio Archaeological Museum.

■ Wheelchair access is limited at Crete's archaeological sites.

Treasures from the Irákleio Archaeological Museum

The famous Agía Triáda sarcophagus

1 Agía Triáda Sarcophagus

Covered with elaborate frescoes that depict daily Cretan life, this limestone sarcophagus is believed to be the only pictorial insight into Minoan funerary rituals. It was discovered at Agía Triáda.

2 Snake Goddess

A ceramic female figure just over 34 cm (13 inches) tall and decorated with fine tin glaze, this faïence statuette is considered one of the finest examples of Minoan art. It was discovered in 1903 at the Palace of Knossos *(see pp30–31)*.

3 The Bees Pendant

Discovered in the Chrysolakkos cemetery close to the Palace at Mália, the Bees pendant dates from the Protopalatial period. It is an important example of gold faïence and depicts two bees carrying honey.

The Bees pendant of Mália

4 The Bull-Leaping Figurine

One of the earliest examples of chryselephantine art, where ivory figures are "clothed" in garments of gold leaf, this figurine from the Palace of Knossos, seen in the leaping position, dates from 1550 BC.

5 The Phaestos Disc

Discovered at Phaestos Palace, this Minoan clay disc is inscribed with 45 pictorial signs that spiral from the outer edge to the centre. It is considered the world's earliest known form of typography.

6 Priest-King Fresco

This fresco depicting the Prince of Lilies, believed to have been the Priest-King of the Palace of Knossos, is brightly coloured with the figure wearing headdress of peacock feathers. It dates from about 1550 BC.

7 Bull's Head Rhyton

This black steatite jug, used for pouring, was known as a rhyton. Fashioned into the head of a bull, the most sacred animal in Minoan culture, it has eyes of rock crystal and eyelashes of jasper.

8 Ring of Minos

Dating back some 3,500 years, this heavy, gold oval-shaped seal ring, believed to have been worn by King Minos himself, depicts a boat sailing between two ports, women working the land and a shrine.

9 Faïence Plaque Collection

A collection of ceramic plaques, which include the famous wild goat and kid found in the sacred treasury rooms of the Palace at Knossos, show the faïence discipline of tin glaze. The plaques are believed to have been inlays.

10 Dolphin Fresco

Depicting lavishly coloured dolphins with other marine creatures, this celebrated late Bronze Age fresco, one of many from Knossos, would have decorated the Queen's Megaron. A replica is now in its place.

MINOAN ART

While little Minoan art remains dating from the Prepalatial period (3500–1900 BC) other than fragments of pottery, there are some magnificent examples from the Protopalatial period (1900–1700 BC) and the Neopalatial era (1700–1425 BC) when Cretans are said to have been encouraged to express themselves creatively. Wall frescoes were painted in vivid colours, with figures and animals depicted in great anatomic detail and showing a natural "movement". Similar detailing was fashioned into gold and silver jewellery, chryselephantine ivory, gold leaf figures and faïence objects. Development of a primitive form of a potters' wheel meant that potters could produce both practical and decorative items of pottery. The Kamares style of symmetrical, rounded vases and decorated black steatite jugs, known as a rhyton, is typical of the period. The Bull's Head Rhyton with rock crystals for eyes, is a prime example. Minoan art often has representations of various sacred animals, such as bulls, snakes and butterflies, as in the ceramic Snake Goddess figurine.

TOP 10 MINOAN ARCHAEOLOGICAL SITES

1 **Palace of Knossos**

2 **City and Palace of Phaestos**

3 **Palace of Maliá**

4 **Palace of Zakros**

5 **City of Górtys**

6 **Royal Villa at Agía Triáda**

7 **Town of Palaókastro**

8 **Sacred Ideon Cave**

9 **Settlement at Myrtós Pýrgos**

10 **Palace of Galatas**

Snake Goddess faïence figurine

Well-preserved ruins, Phaestos Palace

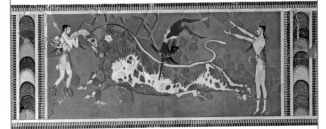

Minoan Bull Leaper fresco, Palace of Knossos

TOP 10 ⭐ Palace of Knossos, Crete

Knossos, the world's largest Minoan site inhabited since Neolithic times, became a powerful commercial and political centre when the legendary King Minos built his palace here in around 1900 BC. The first palace was destroyed in around 1700 BC but this seat of royalty and hub of Minoan life was quickly rebuilt. Discovered in 1878, the second palace comprises a maze of apartments, workrooms and courtyards, many with replica frescoes – originals are in Irákleio Archaeological Museum *(see pp27 and 28)* – showing Bronze Age life.

2 The Palace

Built around a central courtyard, the palace complex **(right)** was arranged with four wings that contained the royal apartments, the throne room, chapels, administration rooms and workshops. Private dwellings dot its periphery. The palace dates back to 2000–1350 BC.

1 Priest-King Fresco

This replica **(above)** of the original brightly coloured fresco of the *Priest-King* wearing a crown of lilies and feathers, is a detail segment from the *Procession* fresco.

3 Hall of the Royal Guard

One of several administrative areas, this hall, which is adjacent to the royal quarters, would have been heavily guarded. It is where the King's guardsmen lived and worked. The wall frescoes displayed here depict shields.

4 Corridor of the Procession

This corridor is the main entrance to the site. It has a copy of the original *Procession* fresco, with over 500 figures.

5 The Throne Room

This imposing room has a carved stone throne copy **(below)**. Arthur Evans believed this was the priest-king's seat but archaeologists now argue the surrounding griffin frescos indicate a priestess sat here.

SIR ARTHUR EVANS

Credited with unearthing the Palace of Knossos, deciphering ancient Cretan script and giving the Minoan civilization its name, Sir Arthur Evans (1851–1941) was a rich British archaeologist. In 1878, Cretan Mínos Kalokairinós had found Knossos but the ruling Ottoman Turks thwarted his excavation. Sir Arthur bought the site in 1898 and began excavation.

6 Workshops

The remains of workshops, used for food preparation and crafts such as pottery, are mainly in the east wing. Excavations revealed many *pithoi*, which were used for storage.

8 The Royal Apartments

These rooms include the King's Megaron (chamber), known as the Hall of the Double Axes, and the Queen's Megaron, which is decorated with a replica of the famous dolphin fresco **(right)** and has an en suite bathroom.

9 Royal Road

Said to be the oldest paved road in Europe, the Royal Road heads northwest of the site to the town of Knossos. The theatre and the Little Palace, a smaller version of the main building, are just off the road.

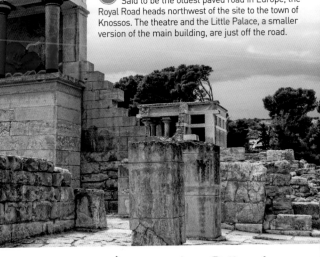

7 Private Houses

Among the remains of private dwellings are the Royal Villa, which is religious in design and probably the home of a priest, and the Villa of Dionysus, where he was cult worshipped.

10 Horn of Consecration

Standing about 3 m (9 ft) tall, this much photographed, restored stone symbol **(left)** is said to represent the horns of a bull, which were considered sacred in Minoan times.

NEED TO KNOW

MAP E6 ■ Knossos, south of Irákleio ■ 2810 231940

Open Apr–Oct: 8am–8pm daily; Nov–Mar: 8:30am–3pm daily

Adm €15, concessions €8; special ticket package available for entry to the Palace of Knossos Archaeological Site and the Irákleio Archaeological Museum *(see p27)* €16, concessions €8

■ Guided tours of the site are available in different languages, along with books on sale at the shop.

■ A café serves light snacks and beverages.

■ The site gets crowded: arrive early before the tourist buses or visit an hour or so before it closes.

⭐ Temple of Aphaia, Égina

Named for princess Aphaia, who became a Greek goddess of fertility, this magnificent temple is thought to have been built around 480 BC with loot that was seized from the Persians after their defeat by the Greeks at the Battle of Salamis. It is located on the site of an earlier temple, the remains of which are said to have been used in its construction. Standing majestically on a pine tree-covered hilltop above the Agía Marína resort, this temple is the most important monument in the Sanctuary of Aphaia.

1 Doric Columns
The temple was built according to the hexastyle format of the Doric period, with 12 outer columns along its longer sides and six along its shorter ends **(above)**. They incline to strengthen the building.

2 The Altar
The remains of Aphaia's altar were discovered at one side of the temple, near the sanctuary's centre. A paved path with a ramp, possibly lined with statues, would have given access to the temple.

3 West Pediment
The re-creation of the west pediment sculpture depicts the mythological hero Ajax **(above)** fighting in the Trojan War. Dating from the Archaic period, the figure is intricately carved in white marble.

4 Sphinx Statues
These statues were positioned at each corner of the temple's roof. With the head and breasts of a woman, body of a lion, a serpent's tail and wings of an eagle, the sphinx was a popular figure of Greek mythology.

7 Prónaos and Opisthódomos
The temple is balanced symmetrically by its portico, *prónaos* (walkways) at its entrance and a small room to its rear, the *opisthódomos*. The cella's façade would have been between these two.

9 Limestone Architraves
Porous local limestone was used to construct the architraves **(above)**, which were plain with a narrow band of plain metopes alternating with carved triglyphs above. They would have been covered with stucco and richly painted.

10 Inner Columns
The inner columns enclose the cella, and are presented in a two-row, two-storey fashion with the lower columns supporting a platform for the upper columns. The fluted design echoes the outer Doric columns.

5 Cella
Traditionally a cella, which is also known as a naos, is an inner chamber housing a cult sculpture of the deity to whom the temple is dedicated. This cella would have had a wooden statue of Aphaia set on a stone base and offerings would have been laid out in front of it.

8 East Pediment
The 480-BC sculpture of the Greek warrior Telamon, which depicts him fighting the Trojan King Laomedon, adorned the temple's east pediment. A re-creation of the sculpture sits there now.

APHAIA
According to legend, Princess Aphaia was saved after she tried to commit suicide by jumping into fishing nets in the waters off Égina, while fleeing from an unwelcome admirer. She was much loved by the goddess Artemis, who, relieved that she had been saved, made her a goddess; a sanctuary was then built in her name. To the Égineans she is their legend, but the Cretans say that the tale took place on Crete.

6 Figurines
Excavated female fertility figurines and pottery **(right)** dating from the late Bronze Age suggest that the temple has been a site of cult worship for more than 4,000 years. Some of these artifacts can be seen in the Archaeological Museum of Égina.

TOP 10 ⭐ Évvia

This long and narrow island, which is separated from mainland Greece by the Euripus Strait, is the second-largest Greek island. Largely mountainous, Évvia was inhabited in prehistoric times when it was home to two wealthy city-states, Chalcis, on the site of its present capital Halkída, and Erétria. The remains of Erétria can be seen near the modern town of the same name. Évvia has a long commercial history and was known for the Euboian scale of weights and measures used throughout Greece. The island has been inhabited by Sicilians, Macedonians, Venetians and Turks, all of whom left their legacy on its architecture and culture.

Évvia's Mountains ①

At 1,745 m (5,720 ft), Mount Dírfys (**right**) is the highest peak on Évvia. It is followed by mounts Óchi, Ólympos, Pixariá and Kandíli, all of which have dramatic rock formations and landscapes of pine forest and grasslands.

Loutrá Edipsoú ②

With its sulphurous warm water (**above**), which has attracted the rich for centuries, Loutrá Edipsoú is the largest spa town in Greece. Neo-Classical buildings line its harbour, while local fishermen continue to ply their trade in the wide bay.

Stení ③

An unspoiled mountain village, Stení lies on the slopes of Mount Dírfys. Affording cooler temperatures than the coastal villages, it is popular with locals and visitors in summer.

Halkída ④

Évvia's capital, Halkída has both modern and historic buildings. Its waterfront boasts many tavernas and hotels. Venetian and Turkish influences can be seen in the Kástro quarter.

NEED TO KNOW

MAP D3

Archaeological Museum of Erétria: 22290 62206; open 8:30am–3pm Tue–Sun; adm €2, concessions €1

■ Évvia can be reached by boat, rail or road. A good network of roads link Halkída with other major towns. Mountain roads can be windy and uneven, so care needs to be taken.

■ Évvia has a selection of eateries. Its tavernas serve traditional food and are a great place to meet locals.

5 Límni

An attractive town of cobbled streets, a pretty harbour and elegant stone houses that hint at its past wealth, Límni **(above)** was a key maritime port in the 19th century. Today, it is a popular holiday spot.

7 Ancient Erétria

Excavations at the site of ancient Erétria began in the late 1890s. The most impressive remains include a theatre. Finds, such as a statue of goddess Athena and pottery and tools dating back to prehistoric times, were unearthed and are in the Archaeological Museum of Erétria.

9 Kárystos

The spectacular slopes of Mount Óchi and the village of Mýli form the backdrop to this small scenic fishing port and holiday hotspot. Its lovely Neo-Classical buildings and Venetian Boúrtzi fortress are worth a visit.

8 Ochthoniá

A Frankish castle and Venetian watchtowers adorn Ochthoniá's skyline. Its coast has a series of deserted beaches that lead to a dramatic wall of cliffs, known as Cape Ochthoniá.

6 Kými

Picturesque, with Neo-Classical mansions overlooking the sea, Kými was a wealthy village during the 19th century due to its thriving maritime industry. Today, it is a place to see traditional crafts and visit the "health-giving" natural springs.

Évvia

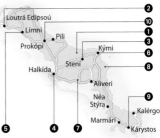

10 Prokópi

Head for Prokópi, a sleepy village on the slopes of Mount Kandíli, for great pine forest views. The village has white-washed stone houses, and pilgrims come to the church **(left)** to worship the 18th-century saint St John the Russian.

The Top 10 of Everything

Boats at Navágio Beach, also known
as Shipwreck Bay, Zákynthos

🔟 Moments in History

① 3000–1000 BC: Cycladic Civilization

The Early Bronze Age civilization that existed in the Cyclades before the advent of the Minoans is thought to be the Greek Islands' earliest. This period is known for its female icons carved from marble. The Cycladic people were believed to be great fishermen.

② 700 BC: Creation of the Dorian Hexapolis

Along with the Ionians and the Aeolians, the Dorians were a tribe of ancient Greece. They created the Dorian Hexapolis, a coalition of six cities. These were Camirus, Lindus and Ialysos on Rhodes, Cnidus and Halicarnassus in Caria, a region of western Anatolia, and Kos.

Cycladic marble figurine

③ 431–404 BC: Peloponnesian War

Unrest between the residents of Corfu and their colonizers, the Corinthians, triggered the Peloponnesian War. Fought between the Athenian and Spartan empires, a peace treaty was signed in 421 BC but the war resumed six years later. The Spartans eventually triumphed at Aegospotami.

Naval battle in the Peloponnesian War

④ 197–146 BC: Roman Invasion

Subjugation by the Romans started in 197 BC and the Roman period began when Corinth was defeated and Greece became part of the Roman Empire in 146 BC. It was a period of great change, and Greece, supported by its isles, became the cultural centre of the Roman Empire. Corinth was rebuilt in 46 BC.

⑤ 1204–1797: Venetian Occupation

The Republic of Venice took control of the Ionians from 1204. This was a key period in the history of the islands – it was due to strong Venetian fortifications that they were able to escape occupation during the Ottoman invasion of Greece. As a result, the islands remained Christian.

⑥ 1309–1522: Knights Hospitallers

Knights Hospitallers, Order of St John of Jerusalem, invaded many islands in the 14th century, particularly those in the Dodecanese. They brought wealth and built strongholds to protect their cities. The Knights were overthrown in Rhodes by the Ottomans in 1522.

7 Ottoman Rule

There have been many periods of Ottoman rule in the islands, the most notable being in the 14th, 16th, 17th and 19th centuries. For example, the Sultan of the Ottoman Empire captured islands in the Cyclades and Dodecanese in the 1500s.

Greek War of Independence

8 1814–1864: British Protectorate

The Greek Islands came under the protectorate of the British in 1814, and, following the Greek War of Independence, overthrew the last period of rule under the Ottoman Empire. The islands acquired Union with Greece in 1864.

9 1941: Axis Occupation

During World War II, the Axis alliance, including Germany and Italy, took control of Greece, and many of the islands were ruled by the Italians. In 1943, the Germans evicted the Italians and exerted their power by sending local Jews to their death.

10 1953: Major Earthquake

The most significant event in modern history was the earthquake that hit the Ionian islands in 1953. This major earthquake, measuring 7.1 on the Richter scale, caused massive damage, destroying many towns.

TOP 10 HISTORICAL FIGURES

1 Hippocrates (460–370 BC)
The ancient Greek physician, known as the "Father of Medicine", was born on Kos. Doctors still take the Hippocratic oath.

2 Alexander the Great (356–323 BC)
Alexander became king at the age of 20. He conquered the Persian Empire and extended his kingdom to India.

3 Emperor Aléxios Komninós (1081–1118)
This emperor ruled during the Byzantine period and was instrumental in building many monasteries.

4 El Greco (1541–1614)
A precursor of Expressionism and Cubism, the paintings of Domenicos Theotokopoulos, *El Greco*, were mostly inspired by religion.

5 Lámbros Katsónis (1752–1804)
An 18th-century naval hero, Lámbros Katsónis fought the Ottomans with his small fleet of pirate vessels.

6 Theodoros Kolokotronis (1770-1843)
Kolokotronis was the pre-eminent leader of the Greek War of Independence (1821–1829).

7 Ioannis Kapodistrias (1776–1831)
The Greek state's first governor after Ottoman occupation in 1827 was from the Ionian island of Corfu.

8 Dionýsios Solomós (1798–1857)
Born in Zákynthos, poet Solomós wrote the Greek national anthem.

9 Eleftherios Venizelos (1864–1936)
A charismatic statesman, Venizelos is known for his promotion of liberal-democratic policies.

10 Maria Callas (1923–1977)
This soprano's musical and dramatic talents led to her being hailed as 'La Divina'.

Statue of Dionýsios Solomós

🔟 Myths and Legends

The ancient hero King Odysseus, portrayed in Homer's great poem

1 Homer's Odyssey

Believed to date from around the 8th century BC, but based on earlier myths, Homer's epic poem *The Odyssey* tells the story of King Odysseus returning home from the battle at Troy as a great warrior to be reunited with his love, Penelope.

2 Poseidon

As per Greek mythology, Paxí was created by the god of the sea, Poseidon, known as Neptune in Latin. Enraged at having no peace and quiet with his wife Amphitrite, he dealt a blow to Corfu with his three-pronged spear, known as a trident, splitting the island in two.

3 Sappho

Legend claims Sappho, a poet from Lésvos, loved Phaon, an ugly ferryman whom the goddess Aphrodite made into a beautiful man. Phaon then rejected Sappho who, broken-hearted, took her life.

4 Minos

The Minoan civilization, which existed from around 2700 to 1450 BC, was named after Minos, the mythical king of Crete. He was the son of Zeus, king of the gods, and lived at Knossos. Minos features on ancient Greek coins.

5 Aegina

In Greek mythology, Aegina was the daughter of river-god Asopus and Metope. After falling in love with her, Zeus, disguised as an eagle, took Aegina to an island and seduced her. She then gave birth to Aeacus, who became the king of the island.

6 The Giant Polyvotis

Polyvotis was a fearsome giant, according to Greek mythology. During a battle with the gods, Polyvotis infuriated Poseidon and the angry sea god cut off a part of Kos and threw it at the giant. The rock, once called Polyvotis, became known as Níssyros.

Statue of Poseidon

7 St John's Vision

Often referred to as John of Pátmos, the saint is said to have written the New Testament's *Book of Revelation* after he saw a vision of Christ in the Cave of the Apocalypse in Hóra, Pátmos. The nearby monastery of St John, founded in 1088 in his honour, is one of the world's most sacred sites (see pp16–17).

8 Helios and Rhode

According to legend, Helios, the sun god, was a handsome Titan who was cult worshipped on an island called Rhode (modern-day Rhodes). The island takes its name from a nymph of the same name with whom Helios had fallen in love.

St Francis of Assisi

9 St Francis of Assisi

The patron saint of animals and nature, St Francis abandoned a life of luxury to devote himself to Christianity. In the 13th century he founded the Moní Theotókou Sisión, near Lourdáta, as a refuge.

10 The Lady of Ro's Flag

Born in 1890 on the Greek island of Kastellórizo, Déspina Achladióti was forced to flee to nearby uninhabited Ro during World War II. She became a hero for raising the Greek flag every day, even though Ro was not a Greek island. The flag is still raised today.

TOP 10 ANCIENT SITES

Ancient sculptures on Delos

1 Delos
Mythological birthplace of Apollo, this sacred island has been inhabited since around 2500 BC (see pp18–19).

2 Pythagóreio, Sámos
The site of an ancient fortified port and known for its 6th-century-BC aqueduct, the Tunnel of Eupalinos (see pp24–5).

3 Phaestos Palace, Crete
The site of a palace dating from 1450 BC considered to be some of the finest Minoan remains (see p27).

4 Ancient Thíra, Santoríni
Ptolemaic, Hellenistic and Roman remains can be seen at the site of the ancient Dorian city of Thíra (see p98).

5 Ancient Akrotíri, Santoríni
This well-preserved, late-Neolithic complex was buried after a volcanic eruption in Minoan times (see p98).

6 Mazarakata Mycenaean Cemetery, Kefalloniá
An important site featuring ancient underground tombs (see p86).

7 Angelókastro, Corfu
MAP A5
This is the site of a 13th-century fortress that formed a hilltop acropolis.

8 Ancient Erétria, Évvia
Dating from the 6th–5th century BC, this ancient city has yielded important Neolithic and Helladic finds (see p35).

9 Paleóchora, Itháki
MAP H2
Once the capital of Itháki, Paleóchora's fascinating medieval ruins include stone houses and churches.

10 Ypapantí, Paxí
MAP B2
Legend has it that this cave was a Byzantine church dedicated to the Presentation of Christ in the temple.

፤10 Museums and Galleries

1 Archaeological Museum Pythagóreio, Sámos

This museum's impressive collection includes pottery dating from the 9th to the 2nd centuries BC. A prize exhibit is a marble statue of Aiakes, father of the tyrant Polykrates, who, in the 5th century BC, seized power in the Aegean and Ionian islands (see p126).

2 Aegean Maritime Museum, Mýkonos

With its collection of nautical instruments, coins, model ships and paintings, this museum preserves Aegean maritime history from Minoan times to the present day (see p94).

3 Antivouniótissa Museum, Corfu

Housed in the 15th-century Church of Panagía Antivouniótissa, this museum is also known as the Byzantine Museum. It has Byzantine and post-Byzantine artifacts and ecclesiastical icons dating from the 15th to the 20th century (see pp12–13).

4 Archaeological Museum, Lefkáda

The artifacts exhibited in this museum (see p75) date from the early Bronze Age. They were excavated from sites around the island, including Kariotes and Nydrí.

5 Irákleio Archaeological Museum, Crete

This fine museum has Minoan artifacts, including faïence figurines and frescoes, ceramics and jewellery (see pp27–8).

6 Museum of Solomós, Zákynthos

With a good collection of furniture, photographs and personal possessions, this museum is dedicated to eminent Zákynthians. These include a number of manuscripts belonging to Dionýsios Solomós, the national poet and author of the Greek national anthem, and the 19th-century writer Andréas Kálvos. The museum houses many portraits of prominent islanders, as well as a collection of coats of arms (see p84).

Ceramic pot, Irákleio Archaeological Museum

Exhibits displayed in Antivouniótissa Museum on Corfu

Mosaic, Rhodes' Archaeological Museum

7 Archaeological Museum, Rhodes

This museum displays Classical, Hellenistic and Roman sculptures as well as mosaic floors from Rhodes Town and Kárpathos. Rhodian funerary slabs dating from the period of the Knights, with personal coats of arms, are also exhibited *(see p14)*.

8 Archaeological Museum, Corfu

Built to house treasures found at the Temple of Artemis on the Mon Repos Estate, this museum now has other exhibits too, including finds from a 13th-century fortress at Kassiópi and the Corfu Town citadel *(see p82)*.

9 Palace of the Grand Masters, Rhodes

Built in the 14th century, the palace was blown up by an accidental explosion in 1856. It was restored in the 1930s and then converted into a museum in 1948. The museum has a fine collection of icons and frescoes, including 12th-century examples painted in the Comneni period *(see p14)*.

10 Helmis Natural History Museum, Zákynthos

MAP H4 ▪ Agía Marina ▪ 26950 65040 ▪ Open May–Oct: 9am–5pm daily; Nov–Apr: 9am–2pm daily ▪ Adm

Exhibits offer an insight into the flora and fauna that make up the island's ecosystem, including rocks, minerals, fossils, corals, fish and birds.

TOP 10 UNUSUAL MUSEUMS

1 Phonograph Museum, Lefkáda
MAP H1 ▪ 26450 21088
This is a private collection of old records and phonographs.

2 Banknote Museum, Corfu
MAP B5 ▪ 26610 41552
Banknotes originating from the German, British, Italian and Greek eras trace Corfu's history.

3 Naval Museum, Zákynthos
MAP H4
Located in Tsilivi, this has exhibits dating from the 1700s to the present day.

4 Municipal Museum of Modern Greek Art, Rhodes
A superb Greek contemporary art collection is on display here *(see p14)*.

5 Folk Museum, Mýkonos
A windmill, plus rare textiles and ceramics, show how rural dwellers lived on the island *(see p94)*.

6 Palaeontological Museum, Tílos
MAP G5
Exhibited are fossils of dwarf elephants that became extinct in 4500 BC.

7 Museum of Contemporary Art, Ándros
MAP N4 ▪ 22820 22444
Works by Matisse, Picasso and local artists are on display here.

8 Kazantzákis Museum, Crete
MAP E6 ▪ 28107 41689
This museum is dedicated to the Cretan writer Níkos Kazantzákis.

9 Vamvakáris Museum, Sýros
MAP N6 ▪ 22813 60914
Musician Márkos Vamvakáris's belongings are displayed here.

10 Museum of Marble Crafts, Tínos
MAP N5 ▪ 22830 31290
Exhibits trace the fascinating history of Tínos marble and how it is crafted.

Museum of Marble Crafts, Tínos

🔟 Monasteries

Dramatically sited Moní Hozoviótissa

1 Moní Hozoviótissa, Amorgós

MAP F4 ■ Near Amorgós Town

An architectural triumph built into the Prophítis Ilías mountainside, this 11th-century monastery was built for the Virgin Mary, protector of Amorgós island. It contains sacred treasures, including a 15th-century icon of the Virgin. Resident monks host a festival here every 21 November.

2 Moní Zoödóchou Pigís, Póros

MAP K3 ■ Póros Town

This 18th-century monastery is called Zoödóchos Pigís (meaning life-giving) because it is built around a curative spring. The white structure is surrounded by pine forest and is a lovely sight. A fine collection of icons and the intricately carved gilded iconostasis separating the nave from the sanctuary are noteworthy.

Moní Panagía of Vlachérna

3 Moní Katharón, Itháki

MAP G2 ■ Anogí

Believed to originate from before the 17th century and built at an altitude of 600 m (1,970 ft), this remote and beautiful monastery is one of the oldest in the islands. It was reno-vated after the earthquake of 1953 and is dedicated to the Virgin Mary.

4 Moní Platytéra, Corfu

MAP A5 ■ Corfu Town

Dedicated to the Virgin Mary and to Saint Chrysanthos, this 18th-century monastery has post-Byzantine icons by famous local painters Klontzás and Ventoúras. It houses the mausoleum of the first Governor of Greece, Ioannis Kapodistrias, a Corfiot who took office in 1827.

5 Moní of the Panagía Odigítria, Lefkáda

MAP H1 ■ Lefkáda Town

Built in the 1400s and noted for its traditional single-aisle architecture, austere exterior and intricate timber roof, this monastery is the island's oldest. It is dedicated to Odigítria, an ancient name for the Virgin Mary depicted in religious icons.

6 Moní Panagía of Vlachérna, Corfu

MAP B5 ■ Kanóni

A landmark of Corfu, this white monastery stands on an islet in the Chalikópoulos Lagoon off Kanóni reached by a causeway. The 17th-century building, which once housed a convent, has interesting architectural features and ecclesiastical icons.

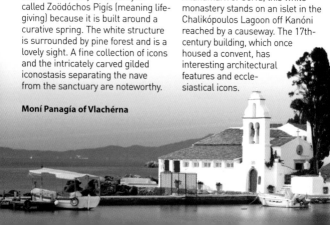

7 Moní of St John of Lagadá, Zákynthos

MAP H4 ■ Katastári

Frescoes and gilded iconostasis dating back centuries feature at this remote monastery. Built in the 16th century and remodelled in the 17th century, it is now home to one monk.

8 Néa Moní, Híos

This austere monastery was built in the 11th century by Byzantine emperor Constantine IX Monomacho and his wife *(see pp20–21)*.

Néa Moní, known for its fine mosaics

9 Moní Transfiguration of the Saviour, Zákynthos

MAP B4

This Byzantine monastery, dedicated to Ágios Dionýsios, the patron saint of Zákynthos, is an isolated structure lying on the island of Stamfáni in the Strofádes, an archipelago that belongs to the Greek Orthodox Church.

10 Moní Agíou Gerásimou, Kefalloniá

MAP G3 ■ Frangáta

A white-walled, red-roofed structure in the Mount Enos foothills, this monastery is dedicated to Gerásimos, Kefalloniá's patron saint. His mausoleum is in the monastery.

TOP 10 RELIGIOUS FESTIVALS

Good Friday festivities

1 Good Friday
Easter, the most important festival of the year, starts with a procession representing the funeral of Christ.

2 Easter Sunday
After midnight every Saturday, worshippers light candles in their local churches to depict the bringing of holy light and good fortune.

3 Annunciation
Marked by a feast, this festival, held on 25 March, is a celebration of when the angel Gabriel announced that Mary would be the Holy Mother.

4 Epiphany
On 6 January, Epiphany events take place throughout the islands and are marked by the blessing of water.

5 Lenten Monday
Religious services, feasts and carnivals are held in the three weeks prior to Lenten Monday (Katharí Deftéra).

6 Lent
A period of penance to commemorate Jesus fasting before Easter, Lent culminates on Easter Sunday in April.

7 Pentecost
Sometimes known as the Feast of Weeks, Pentecost marks the descent of the Holy Ghost on the Apostles.

8 Dormition of the Virgin
The most important religious celebration after Easter takes place on 15 August throughout the islands.

9 Traditional Festivals
Taking place throughout the year, these festivals *(panegíria)* celebrate each island's patron saint.

10 Christmas
Christ's birth is celebrated with services, prayers and the giving of gifts.

🔟 Beach Resorts

① Blue Lagoon Resort, Kos

A theatre, hair studio, crêpe stand and teenagers' dance bar are some of the features of this family resort with 600 guest rooms. It also has swimming pools, sun terraces and restaurants, plus a children's centre. Votsalo Spa is the perfect place to relax, offering a variety of beauty treatments, as well as a gym (see p171).

Swimming pool, Blue Lagoon Resort

② Negroponte Resort, Evvia

Built in stone and painted with Mediterranean colours to blend with its surroundings, this five-star resort offers 100 top-quality guest rooms. As well as standard rooms, guests can book family rooms and luxurious suites, which are spacious and stylish. All rooms have amenities such as internet access. There is a restaurant, children's play-ground and activity programme, plus swimming pools and tennis courts (see p175).

③ Mediterranean Beach Resort, Zákynthos

A luxurious resort comprising eight buildings designed to echo a traditional Ionian village, the Mediterranean Beach has 113 rooms, maisonettes and suites. The accommodation is situated in gardens around a swimming pool and over-looks the bay. On-site amenities include a gym, restaurant, bars and a children's area (see p168).

④ Lemnos Village Resort, Límnos

Among the facilities at this sprawling five-star resort are tennis courts, watersports and swimming pools, along with a choice of restaurants. Children have their own pool and play area. Offering 134 stylish rooms and suites, this resort lies next to the beach (see p173).

⑤ Archipelagos Resort, Páros

A compact five-star beach resort that is characterized by its Cycladic-style square white buildings, the Archipelagos has a modern feel. The guest rooms are painted and decorated in white with splashes of vibrant colour. The poolside terraces, Greek à la carte restaurant and fitness suite are also elegant (see p169).

⑥ Esperos Village Resort, Rhodes

This luxurious adults-only resort has stunning sea views and offers pleasing features such as guest rooms with a personal bar area. Activities, including tennis courts, a fitness suite, a spa and water-sports, keep guests entertained. There is also a piano bar and an excellent à la carte restaurant specializing in Mediterranean dishes (see p171).

7 AKS Annabelle Village, Crete

One of Crete's foremost family beach resorts, the five-star AKS Annabelle offers bungalow-style suites with sea and garden views. It overlooks a Blue Flag beach. Services for children include play areas, paddle pools, the Cinderella Club, a special buffet menu and babysitting services *(see p178)*.

Suites at AKS Annabelle Village

8 Candia Maris Resort & Spa, Crete

The 285 rooms and suites of this beachside resort are all beautifully appointed and offer internet access. The hotel is spread out in six separate buildings, surrounded by gardens. Deluxe rooms have a separate pool. The spacious seawater spa has many thalasso-therapy treatments. The three excellent restaurants serve buffet and à la carte dining *(see p178)*.

9 Paradise Beach Resort and Camping, Mýkonos

Popular with visitors who like to party, this beachside resort and campsite has facilities like canoeing, water- and jet-skiing by day and dancing by night. The site also offers apartments and cabins and has a restaurant and internet café on site *(see p170)*.

10 Santoríni Kastelli Resort

This deluxe resort reflects Aegean Cycladic architecture with its white-washed walls and splashes of colour. It has an elegantly decorated lounge and conference areas, bars and restaurants, along with tennis courts and swimming pools. Its spa has a steam room and provides massage and aromatherapy. The resort is a short walk from the black volcanic sands of Kamari beach *(see pp170–71)*.

The pool at Santoríni Kastelli Resort

🔟 Bays and Beaches

Views out over the bay and Navágio Beach, enclosed by white cliffs

1 Navágio Beach, Zákynthos

Also known as Shipwreck Bay because of a rusty freighter that sits partially buried in sand, this white-sand beach lies in a sheltered bay enclosed by soaring cliffs. Although world famous, it is fairly quiet here, except when day-trippers arrive to take photographs. It is signposted from Volímes village on the island's northwest coast (see p84).

2 Falásarna Beach, Crete

Widely considered to be one of the finest beaches on Crete, this pretty but often windy spot has golden sand and shallow turquoise waters. It lines one of the bays of the Gramvoússa peninsula and has a small harbour. Nearby are the ruins of the ancient Falásarna city (see p105).

3 Yaliskári Beach, Corfu
MAP A5

This small, sandy beach, boasting clusters of rock formations and a pine tree forest that descends almost to the water's edge in places, is one of the most secluded beaches along the west coastline. It lies in a lovely bay with crystal clear waters and is, unsurprisingly, a popular picnic and swimming spot with locals.

4 Ágios Geórgios Beach, Santoríni

Consisting of the award-winning Blue Flag Périssa and neighbouring Perivolos beaches, the long stretch of sand at Ágios Geórgios is touristy, but nonetheless picturesque, with the black sand and clear waters giving it its identity (see p92).

5 Banana Beach, Skiáthos

One of the two banana-shaped halves of the beach is picturesque and sandy. It has umbrellas, sun-loungers, the odd taverna and a few businesses offering watersports. The other half, locally known as Petit Banana, is quiet, secluded and popular with naturists (see p139).

Sunbeds and shade on Banana Beach

6 Eliá and Agrári Beaches, Mýkonos

The long Eliá and Agrári beaches merge almost seamlessly, except for a cluster of rocks, to form the island's longest stretch of sand. Eliá, a Blue Flag beach, is more organized, with parasols and cafés, while Agrári is quieter and more relaxing *(see p92)*.

7 Stáfylos Beach, Skópelos

Located just 4 km (2 miles) from Skópelos Town and reached by a staircase from the road, this small pretty beach is popular with locals and tourists alike. It is surrounded by pine-covered hills and has crystal clear waters *(see p138)*.

8 Fínikas Bay, Sýros

A natural harbour sheltered from the northern *meltemi* winds by a rocky backdrop, this pretty bay is always buzzing with activity from fishing boats. It is the island's second port after Ermoúpoli *(see p92)*.

Vivid blue sea at Vaï beach

9 Vaï Beach, Crete

Backed by a forest of centuries-old native Cretan date palms, this large crescent-shaped beach, with its golden sand and dazzling blue water, is one of the most popular spots on the island *(see p105)*.

10 Lithí Beach, Híos
MAP K5

This huge horseshoe-shaped beach, just west of Híos Town, is famous for its natural beauty, its astonishing sunsets and the many fine fish tavernas that vie for attention.

TOP 10 DIVING HOTSPOTS

Divers exploring the sea, Crete

1 Chaniá Bay, Crete
MAP D6
Shipwrecks and amphorae are visible in the shallow waters of this bay.

2 Fiskárdo, Kefalloniá
Dive sites off the coast of this pretty village *(see p86)* are known for their underwater life, including damselfish.

3 Kalafátis, Mýkonos
MAP P6
Shoals of barracuda swim amongst shipwrecks off this beach, including a wreck 20 m (66 ft) in the sea.

4 Ágios Nikólaos, Crete
See ancient remains, wrecks, caverns and reefs at this popular diving spot near a holiday resort town *(see p104)*.

5 Psalídi, Kos
MAP Y1
Psalídi is known for its sheer cliffs, caverns full of sponges and sea bream.

6 Laganás Bay, Zákynthos
Between April and June loggerhead turtles come here to breed *(see p84)*.

7 Périssa, Santoríni
MAP V3
Fascinating underwater rock formations created by the lava from this volcanic island can be seen off Périssa.

8 Xirókampos, Léros
MAP F4
Ancient artifacts are among the things to see along this stretch of coast.

9 Paleokastrítsa, Corfu
MAP A5
Caves and marine life draw divers to Liapádes Reef, off Paleokastrítsa.

10 Vasilikí, Lefkáda
MAP G1
Many dive sites can be found off Vasilikí, including canyons and cave swims.

🔟 Secluded Beaches

① Ágia Marína Beach, Spétses

Spétses has some of the best beaches in the Argo-Saronics, with Ágia Marína beach being one of the most popular. It is overlooked by a tiny pink and white chapel of the same name. For seclusion, head off around the headland, between some rocks, to a tiny bay of fine golden sand and pebbles lapped by the sea. Excavations here have unearthed Early Bronze Age remains *(see p149)*.

② Psilí Ámmos Beach, Pátmos

Tucked cosily in a scenic cove, Psilí Ámmos, meaning "fine sand", lives up to its name with soft golden dunes dotted with tamarisk trees, which provide visitors with welcome shade. Access to the beach is on foot along a small rocky track or by fishing boat from Skála, which means it is rarely crowded *(see p114)*. A small taverna at the north end serves local cuisine.

Well-concealed Psilí Ámmos beach

③ Megálo Limonári Beach, Meganíssi

This long peaceful stretch of beach, which is shielded by trees, is accessible only on foot or bike along a track from Katoméri village. As a result, only those keen on seclusion are likely to make the journey here. However, it is well worth the effort as the sandy beach is beautiful and the crystal-clear water is idyllic for swimming *(see p78)*.

Cliffs framing Pórto Katsíki beach

④ Pórto Katsíki Beach, Lefkáda

Reached by descending a series of steps from the parking area, or by boat from the resorts of Nydrí and Vasilikí, this beach attracts visitors because of its reputation as the island's most beautiful spot. It has fine golden sand, striking white cliffs topped by lush vegetation and some interesting rock formations *(see p78)*.

⑤ Agrári Beach, Mýkonos

Sand dunes set amongst green shrubs characterize this secluded beach, which is popular with nature lovers. It is enclosed in a bay by a hillside of forest and rock formations, and can be reached by a sharply descending narrow footpath or by caïque. It has a small taverna. Nearby is the busier Eliá beach *(see p92)*.

⑥ Pahiá Ámmos Beach, Tínos

Pahiá Ámmos beach is a world away from the crowds: the only noise you will hear is the sound of the sea breaking gently on the shore. This serene spot is surrounded by rugged terrain, with a small road providing access. Set in a natural bay, the white beach is made up of a series of sand dunes *(see p92)*.

7 Adrína Beach, Skópelos

Reached from Pánormos by walking through a forest of scented pine trees, or by boat, this peaceful beach is often deserted. It is a mix of pebbles and sand, with lots of tiny coves to explore. Opposite is Dasiá island, which, according to legend, is named after a female pirate who drowned off its shore *(see p139)*.

8 Ftenágia Beach, Chálki
MAP G5 ■ Near Emporió

A quintessential secluded beach, Ftenágia is nestled in a small cove along a hiking route from Emporió village, which is its only access. The beach is popular with yacht owners, who often anchor offshore so they can go for a swim in its safe waters. The beach is a mix of sand and pebbles, and has a small taverna.

9 Elafoníssi Beach, Crete
MAP D6 ■ Elafoníssi islet, off Palaiochóra

With its white sand that has a pinkish tint, due to fragments of coral, and its great expanse of shallow turquoise sea, this beach is one of the most idyllic spots on Crete. The beach is located at the far south-eastern tip of the island on the Elafoníssi islet and can be reached by crossing the reef on foot or by boat. For more active types, kiteboarding and windsurfing are available and there are a few local tavernas nearby.

Horseshoe-shaped Kouloura beach

10 Kouloura Beach, Corfu
MAP B5 ■ Near Kalámi

Set in a tiny attractive cove, which is surrounded by cypress trees that create a feeling of total privacy, this sandy, horseshoe-shaped beach has escaped development and offers a peaceful place to relax. Its plentiful trees provide ample shade. The beach looks out over the picturesque fishing harbour of Kouloura village, where you will find a handful of small tavernas serving refreshments.

Grassy sand dunes, Elafoníssi beach

TOP 10 Picturesque Villages

Charming houses clustering around the harbour at Fiskárdo

1 Fiskárdo, Kefalloniá

This picture-postcard village retains many original Venetian buildings. Painted in pastel shades, the houses that were once the homes of merchants line its yacht-filled harbour. Most now host tavernas, restaurants and shops *(see p86)*.

2 Imerovígli, Santoríni

With dazzling blue and white square houses and blue-domed churches descending down the hillside to the bay, Imerovígli is a typical Santoríni village. Facing west, over the flooded caldera, its sunsets are a photographer's dream *(see p98)*.

Whitewashed houses, Imerovígli

3 Pýrgos, Tínos
MAP N5

Buildings made of white marble give this village a distinct beauty. Marble has been used in the area for centuries, making it the birthplace of many great sculptors and home of the Tínos School of Fine Arts.

4 Náoussa, Páros
MAP E4

Considered one of the prettiest villages in the Cyclades, this small fishing port is characterized by white houses, chapels, a bustling harbour and tiny lanes.

5 Anogí, Itháki
MAP G2

A village of olive trees and stone walls, Anogí is about 500 m (1,640 ft) above sea level. One of Itháki's most remote communities dating from medieval times, the people observe old customs and speak a distinct dialect.

6 Lagopóda, Zákynthos
MAP H4

This mountain village offers a rare glimpse into how the people of Zákynthos have lived for centuries, and is home to cobbled alleyways, stone houses and a pretty church.

7 Pélekas, Corfu
MAP A5

This picturesque village has resisted mass tourism. Bougainvillea-covered stone houses sit among olive groves and vineyards, small hotels and traditional tavernas. Sitting up on a hill overlooking pretty, unspoiled sandy beaches, Pélekas is known for its breathtaking sunsets.

8 Kontiás, Límnos
MAP E1

Kontiás is one of the most charming villages on the island. Its delightful architecture reflects Byzantine, Venetian and Ottoman influences. Surrounded by pine forest, it has some glorious beaches. Kontiás is the seat of the municipality.

9 Asfendioú Villages, Kos
MAP Y2

Comprising Ziá (Evangelístria), Ágios Dimítrios, Asómatos and Lagoúdi, this cluster of villages on the slopes of Mount Dikéos is surrounded by forest and springs. The villages retain their old charm with stone houses and Byzantine churches.

Stunningly picturesque, Oía

10 Oía, Santoríni

A popular holiday spot, Oía is a lovely village of blue-domed churches, white houses and tiny alleyways huddled on the hillside overlooking the volcano. It has many tavernas, gift shops and the remains of a Venetian fortress *(see p98)*.

TOP 10 TREES AND FRUITS

Fig trees heavy with fruit

1 Figs
The deep purple fig can be seen growing amongst the large leaves of fig trees in summer.

2 Olives
Greece is home to the largest variety of olives in the world. The green and black olives found in groves throughout the islands are either eaten or made into oil.

3 Vines
Vines grow wild or in vineyards, with the green and black grapes used to make wine and liquors.

4 Apples
Local apples are usually small and sweet, and are used for cooking or served at the end of a meal.

5 Pomegranates
Pomegranates, which can be seen growing in late summer, were mentioned in Homer's *Odyssey* and are used to make grenadine.

6 Bananas
The local variety of this popular fruit is small and grown on the lower, warmer plains of the islands.

7 Cretan Plane Tree
Endemic to Crete, the striking plane tree is often found in villages and provides welcome shade.

8 Carob Tree
With its dark seed pods used as a chocolate alternative, the carob tree is a key part of the landscape.

9 Cypress Tree
Tall and elegant, the cypress tree is from the conifer family and is seen in gardens and in the wild.

10 Cedar Tree
Forests of the broad-branched cedar tree can be found at higher altitudes throughout the islands.

🔟 Natural Wonders

1 Extinct Volcano Polyvótis, Níssyros

MAP X3

One of Níssyros's main attractions, this volcano is around 260 m (860 ft) in diameter and 30 m (100 ft) deep, and surrounded by a grey landscape of craters. Steps lead down to its core. Its last eruption is said to have been over 5,000 years ago.

Volcano crater, Polyvótis

2 Islets of the Greek Islands

Islets dot the sea around the Greek Islands. Some, such as Néa Kaméni off Santoríni, are uninhabited and volcanic, while others, such as the Diapontian islands off Corfu, have rich vegetation, small communities and some fantastic beaches.

3 Blue Caves, Zákynthos

A rock formation of arches and caverns, the Blue Caves are among the most popular natural wonders of Zákynthos. The sea appears bright blue due to the reflection of the sun between the arches, and contrasts sharply with the white cliffs *(see p84)*.

View of the Blue Caves from out at sea

4 Samariá Gorge, Crete

A stunning creation of nature, the gorge stretches about 16 km (10 miles). Towering rock walls have been created over millennia by a river, and at a spot called Iron Gates (Sideróportes) they are only 3 m (10 ft) apart *(see pp27 and 103)*.

5 Korissíon Lagoon, Corfu

This freshwater lagoon is separated from the sea by a narrow stretch of sand dunes. Enclosed by a forest, it is a natural habitat for wildlife and many bird species, including sandpipers and egrets *(see p82)*.

6 Thérmes Hot Springs, Kos

MAP Y2

These are perhaps the most famous hot springs on the Greek Islands. Thérmes beach *(see p114)* is known for its naturally warm pool in the rocks, which is fed by hot springs and is said to have curative powers.

7 Nydrí Waterfalls, Lefkáda

These waterfalls, located in the Dimosári gorge, are the most spectacular of the many falls on the island. Cascades of water fall into pools of crystal-clear water, which are popular with swimmers *(see p77)*.

8 Mýrtos Bay, Kefaloniá

This highly photographed bay is famous for its dazzling white-pebble beach formed by fragments from the calcite-rich limestone cliffs that surround it, and its vivid turquoise sea. Mýrtos beach has been voted one of the best in the world *(see pp78 and 86)*.

Boat tours entering Melissáni cave

9 Caves of Sámi, Kefalloniá

Melissáni cave and lake are the highlight of an extraordinary network of subterranean waterways and caverns originating at Katavothres, on the other side of the island. It is said to have been the sanctuary of the god Pan and several nymphs, including Melissáni. Nearby is the Drogaráti cave with its famous stalactites and stalagmites (see p76).

10 Caldera, Santoríni
MAP U2

This water-filled cauldron, enclosed on three sides by 300-m- (980-ft-) high cliffs, lies off Santoríni and was created in 1450 BC, when a huge volcanic eruption blasted through the once-circular island, creating its current crescent shape. At the centre of the caldera, the Néa Kaméni and Palaia Kaméni islands emerged from the water with subsequent volcanic activity. The volcano remains active.

TOP 10 ANIMALS AND BIRDS

1 Monk Seal
One of the rarest seals in the world, the monk seal can be seen off the islands, in particular in the Cyclades and northern Sporades.

2 Moorish Gecko
Usually found "sunbathing" on walls in summer, this grey-brown reptile can grow to around 15 cm (6 in) in length.

3 Mílos Viper
This venomous snake is endangered, but found mainly on the Mílos and Kímolos islands in the Cyclades.

4 Freyer's Grayling Butterfly
Small and dark brown in colour, the Freyer's Grayling is native to the Mediterranean and normally seen in July and August.

5 Kri-Kri
A protected species, this wild goat is a native of Crete and is mainly found roaming in the Samariá Gorge (see p103).

6 Dolphin
Look out for bottlenosed, striped and common dolphins when sailing in Ionian and Aegean waters.

7 Sardinian Warbler
This striking native Ionian bird can be identified by its black head, pale underside and red eye ring.

8 Scops Owl
These small grey owls are hardly visible but their trill can be heard at night.

9 Eleonora's Falcon
Kestrels, eagles and falcons, including Eleonora's falcon, are seen in large numbers around the islands.

10 Loggerhead Turtle
Zákynthos' Laganás Bay (see p84) is one of the key breeding grounds for this endangered species of turtle.

A loggerhead turtle

TOP 10 Outdoor Activities

1 Watersports
Plaka Watersports: www.plaka-watersports.com

Snorkelling and dive sites are found on almost every island, with qualified companies offering tuition and outings for beginners, as well as challenging deep dives. Surface watersports like windsurfing, which is popular in Náxos because of the strong winds, along with water-skiing and kayaking, are available. Families can also enjoy the thrill of banana rides and pedalos.

2 Horse Riding
Naxos Horse Riding Club: www.naxoshorseriding.com ■ **Trailriders Club: www.trailriderscorfu.com**

Privately owned riding centres are dotted around the islands, including the Náxos Horse Riding Club on Náxos and the Trailriders Horse Riding Club on Corfu. Taking a trek through olive groves, farmland, woodland, mountains or along the beach in the early morning or late evening is a magical experience.

Golfer putting on the green, Crete

3 Golf
Hellenic Golf Federation: www.hgf.gr

Although there are only a handful of international-standard courses in the Greek Islands, golf is popular on the islands. Among the best courses are the 18-hole ones at Ropa Valley on Corfu, Afandou Golf Course on Rhodes and Crete Golf Club at Chersónissos (Hersonissos), Crete. The sport is promoted by the Hellenic Golf Federation.

Tourists hiking in the Samariá Gorge

4 Walking
Strata Tours: 28220 24249; www.stratatours.com

The islands offer fabulous country-side to explore on foot, with some areas, such as Crete's Samariá Gorge *(see p103)*, having designated hiking trails. These pass by valleys and rivers, villages, vineyards and olive groves, plus the odd taverna for refreshment. Always carry bottled water and a phone.

5 Tennis
Hellenic Tennis Federation: 21075 63170; www.efoa.org.gr

Most of the larger hotels and resorts have courts, and municipality courts are generally open for visitors' use. Tennis has been played in Greece for centuries; in fact, it was a feature of the celebrations marking the first modern Olympic Games in Athens in 1896. The regulating body is the Hellenic Tennis Federation.

6 Paragliding
Lefkada Paragliding: www.lefkadaparagliding.gr

Major resorts such as Kos and Rhodes in the Dodecanese, Mýkonos and Páros in the Cyclades, Nydrí on

Lefkáda and Sidári on Corfu offer the chance to experience paragliding. Charges are usually for an hour.

7 Adrenaline Activities
Lefkada Paragliding: www. lefkadaparagliding.gr

All the islands' main holiday resorts offer adrenaline sports for those who love the idea of jumping off a bridge on the end of a bungee cord, kayaking down a river, parachuting from a plane, canoeing, depth diving or taking a kamikaze water ride at a theme park.

8 Fishing
The islands are known for their fishing harbours. While most of these are industry-driven to provide fish and seafood for restaurants, others offer organized pleasure trips. Among the harbours where boats leave for trips are Égina in the Argo-Saronics, Pátmos in the Dodecanese and Fiskárdo in Kefalloniá.

9 Sailing
Greek Sails: www.greeksails.com

The Greek Islands boast some of the best sailing waters in the world, as well as marinas and anchorages. Although tidal variation is minimal and the climate is mild, winds can vary between the island groups and present a challenge. For example, in the Ionians, the *maïstro* wind tends to be gentler than the Aegean Sea's *meltemi*, which can reach gale force.

Cyclists in the White Mountains, Crete

10 Cycling
Santorini Adventures: www. santoriniadventures.gr

For experienced cyclists, cycling around the Greek Islands can be very rewarding and it is great way of getting around. Bicycles, mountain bikes and mopeds can be hired at moderate cost in most resorts, including those on Santoríni and Mýkonos in the Cyclades, Corfu or Crete. Enjoy leisurely family outings along scenic plains and coastal roads. For endurance cyclists, the rugged hilly terrain of many of the islands can be a challenge.

Sailing dinghies pick up the wind of the stunning coastline of Spétses

🔟 Children's Attractions

Fun water attractions and slides suitable for kids, Aqualand on Corfu

① Aqualand, Corfu
MAP A5 ▪ Ágios Ioánnis
▪ 26610 58531 ▪ Open May–Oct:
10am–6pm daily ▪ Adm ▪ www.
aqualand-corfu.com

An amazing collection of water rides, huge slides and other water-based attractions, including a giant spa pool, the Black Hole, the Kamikaze and Crazy River, make Aqualand an oasis of fun in the countryside.

② Sun Cruises Glass-Bottom Boat, Kefalloniá
MAP G3 ▪ Port of Argostoli ▪ 26710 25775 ▪ Open from 9am daily ▪ Adm ▪ www.captainmakis.gr

This glass-bottom boat sails to the island of Agios Nicholas to see an

Glass-bottom boat, Kefalloniá

ancient shipwreck, and on to the uninhabited Vardiani Island before going ashore to Xi beach for a barbecue. Dolphins often swim alongside the boat, adding a delightful bonus.

③ Acqua Plus Water Park, Crete
MAP E6 ▪ 5th km Hersonissos to Kasteli ▪ 28970 24950 ▪ May–Oct: 10am–6pm daily (Jun & Sep: to 6:30pm; Jul & Aug: to 7pm) ▪ Adm ▪ www.acquaplus.gr

One of the most popular waterparks on Crete, Acqua Plus has many attractions, such as Kamikaze Hydrotube, Acquaslalom, Black Holes, a Crazy River and a slide under a weeping willow tree. There is also a children's section.

④ Bowl'm Over, Zákynthos
MAP H4 ▪ Tsiliví ▪ 26950 25142 ▪ Open 11am–late daily ▪ Adm ▪ www.bowlmover.gr

With four bowling lanes, a pool, foosball and air-hockey tables and a children's play area, this is a great place to have some holiday fun with the family. Snacks and drinks can be enjoyed on the terrace or in the gardens.

⑤ Hydropolis, Corfu

MAP A4 ■ Acharávi ■ 26630 64000 ■ Open May–Sep: 10:30am–6pm daily ■ www.gelinavillage.gr

Set in a large estate of landscaped gardens in northern Corfu, Hydropolis combines a leisure and sports centre with numerous swimming and water-volleyball pools and giant slides.

⑥ Luna Fun Park, Zákynthos

MAP H4 ■ Tsiliví ■ 26950 48035 ■ Open 10am–late daily ■ Adm ■ www.lunafunpark.com

The laser arena here is designed like a war zone. A retired World War II aeroplane and a climbing tower are a few of the star attractions. A baby sitting service and refreshments are also available.

⑦ Aquarium, Rhodes

MAP V4 ■ Kos Street, Rhodes Town ■ 22410 27308 ■ Open Apr–Oct: 9am–8:30pm daily; Nov–Mar: 9am–4:30pm daily ■ Adm ■ www. rhodes-aquarium.hcmr.gr

A centre for marine research and a hospital for injured marine life, this fascinating aquarium is also a tourist attraction. Dolphins, sharks, seals, sea turtles, fish and crabs can be seen in the tanks.

Shark tank at the CretAquarium

⑧ CretAquarium, Crete

MAP E6 ■ Gournés Naval Base, Irákleio ■ 28103 37788 ■ Open May–Sep: 9:30am–9pm daily; Oct–Apr: 9:30am–5pm daily ■ Adm ■ www. cretaquarium.gr

There are around 2,500 sea creatures representing 200 Mediterranean species of marine life at this state-of-the-art aquarium, with viewing tanks, multimedia presentations and observation points.

⑨ Water Park, Santoríni

MAP V3 ■ Périssa ■ 22860 83311 ■ Open 10am–7pm daily ■ Adm

Family fun is assured here with swimming pools, slides, flumes, a restaurant with a lovely dining terrace, a gift shop and sunbathing areas. A children's playground features its own pool and a large-scale toy galleon.

⑩ Lido Water Park, Kos

MAP X2 ■ Mastichári ■ 22420 59241 ■ Open 10am–6:30pm daily ■ Adm ■ www.lidowaterpark.com

This huge waterpark complex offers swimming pools for adults and kids, a large wave pool, a giant spa, and thrilling slides and flumes.

Entrance to the Aquarium, Rhodes

Nightlife Spots

1 DiZi Bar, Corfu

By day this bar serves coffees and cocktails, and then transforms into one of Corfu's liveliest evening venues as the sun goes down. Music from indie and hip-hop to serious rock adds to the party atmosphere, plus there are live music events and themed reggae or Greek nights (see p80).

2 Skandinavian Bar, Mýkonos

Open since 1978, this bar is an established fixture on the club scene and is always popular. Famous DJs, good music and a variety of groovy cocktails make for a memorable night. The dancefloor is upstairs, but there's also a great patio outside, which is perfect for cooling off (see p95).

Cocktails at popular DiZi Bar

3 Mylos Bar & Restaurant, Santoríni

Located on the terrace of a brilliant white windmill right on the edge of Santoríni's caldera, and serving gourmet appetizers through to desserts, an evening spent at

Mylos Bar will be an unforgettable experience. It offers a serious wine list and cocktails presented with flair too (see p99).

4 Zig Zag Bar, Rhodes

A landmark venue housed in a stone building with decorative arches, Zig Zag has become something of an institution in Pèfkoi. The atmosphere is always buzzing and the staff are friendly. It serves local drinks like ouzo, cocktails and beers, as well as an impressive wine selection by the glass (see p117.

5 Bass Club, Kefalloniá

Bass Club has become known for its lively music nights when DJs and local bands play anything from Greek and foreign rock to R&B tunes. It's every inch a cosmopolitan party venue. Along with cocktails, its kitchen serves breakfast, lunch and dinner, plus a good range of café-style snacks (see p80).

6 Jolly Roger, Crete

Right by the beach and overlooking the sea, the Jolly Roger is a lively place decked out as a pirate boat. The mix of locals and visitors who frequent this bar every night gives the Jolly Roger an unbeatable atmosphere. It offers live music, quiz nights and a whole host of cocktails (see p107).

7 Delon Pub, Kos

Run by a friendly Greek family, this popular bar has customers returning year after year for the first-rate beer and reasonable prices. The bar has an English vibe, so do not come here for an authentic Greek experience. Sports fans are catered for with big-screen TVs, although one of these is usually given over to UK soaps. The English breakfasts come highly rated (see p121).

The popular Mylos Bar & Restaurant

8 Rescue Club, Zákynthos

This is the biggest club on Zákynthos, and also on the Ionian islands, with a capacity for over 2,000 revellers. There are three huge rooms and six bars, which host world-famous DJs and some unusual theme nights. These include a paint party and "Back to School" disco. Rescue Club underwent a massive revamp in 2017 and promises even bigger and better parties than before (see p80).

Rescue Club packed to the rafters

9 Palazzo Cocktail Bar, Corfu

This bar, located in the centre of Sidári, serves breakfast, coffee and international meals by day, then transforms into one of Sidári's liveliest evening venues. The resident DJ plays music ranging from indie to hip-hop, and on occasion is accompanied by live music. Like many other venues in Sidári, it unashamedly caters for tourists, with darts and a pool table, and sport shown on large TV screens. There is an outdoor seating area and a kids' play area too (see p80).

10 Socratous Garden, Rhodes

Surrounded by palm trees and banana plants, this garden venue is an informal place to spend time. By day it serves light snacks and coffees shaded from the sun under trees, and by evening becomes an atmospheric place with a menu of Greek dishes, accompanied by colourful cocktails and music (see p117).

TOP 10 DRINKS

1 Retsína
A powerful white (or rosé) wine that is flavoured with pine resin, *retsína* is the traditional wine of Greece. It is quite often served chilled.

2 Greek Coffee
Brewed in a copper pot known as a *bríki*, thick and strong Greek coffee is produced by boiling finely ground coffee and sugar together.

3 Metaxa
Metaxa is a famous brand of strong, cask-matured wine. It is dark brown in colour and tastes rather like brandy.

4 Tsípouro
Tsípouro is distilled from the residue of grapes as well as olives, and is a popular after-dinner liqueur.

5 Cocktails
Cocktails are mainly served in beach bars and are almost always colourful and highly decorated.

6 Kumquat Liqueur
A popular speciality on Corfu, this is a sweet liqueur that is flavoured with the citrus fruit kumquat.

7 Mythos Beer
This light lager is probably the best-known of all the Greek beers.

8 Frappé
A great way to enjoy coffee when the weather is hot, a refreshing frappé is a frothy, iced beverage.

9 Soft Drinks
Major international soft drinks are sold in the Greek Islands, and there are some good local brands too.

10 Oúzo
Oúzo is a strong, aniseed-flavoured spirit, which can be enjoyed neat or mixed with water, which turns it cloudy.

Oúzo, spirit mixed with water

TOP10 Greek Dishes

A hearty plate of *kléftiko*

great way to sample many local dishes in one meal. Typical main dishes include octopus, shrimp and *keftédes*, *souvláki* and *kléftiko* for the meat option. Most restaurants also offer vegetarian *mezédes*.

4 Loukoumádes
A sweet snack, often served at street festivals, *loukoumádes* are small dough balls deep-fried in olive oil and sprinkled with sugar. Many restaurants also dip the balls in honey or sugar syrup, sprinkle with cinnamon and serve with ice cream.

5 Souvláki
Souvláki is a dish of diced pork on skewers, chargrilled and served with lemon and herbs, accompanied by salad and fried potatoes or pitta bread. It is also a popular fast-food option, served in *souvlatzídiko* outlets in pitta bread with salad and dressing. Chicken, fish and vegetables are modern variations.

1 Kléftiko
Kléftiko, meaning stolen meat, is a dish of lamb or goat that is cooked slowly over coals in a special oven, sometimes for around 8 hours. Traditionally the meat would be sealed in paper with herbs to keep it moist and flavoursome, but nowadays foil is more commonly used.

6 Moussakás
A delicious dish made of aubergine sautéed in olive oil, sliced tomatoes and potato layered with minced meat – usually lamb – and topped with a béchamel sauce. Lots of garlic, herbs, spices and onion, plus a dash of wine, give it its flavour. The dish is generally served with a salad.

***Souvláki*, served on pitta bread**

2 Glyká Koutalioú
These confectioneries can be eaten as a snack or to complete a meal. Cherries, pineapple pieces, melons and even peppers, aubergines, almonds and walnuts are candied until they are soft and syrupy. *Glyká koutalioú*, literally "spoon sweets", have been made in the villages for decades and are often served with ice cream.

3 Mezédes
An assortment of small dishes, usually starting with salad, dips and pitta bread, *mezédes* are a

***Moussakás*, a popular Greek dish**

7 Pastitsáda

A speciality on Corfu – where it is a traditional Sunday dinner – but served throughout the islands, *pastitsáda* is a dish of chicken, or occasionally beef, cooked in the oven with tomato, peppers, onion and cinnamon. It is generally served with pasta. Variations include substituting the meat with seafood such as lobster, or using tofu.

8 Keftédes

These tasty meatballs, which are made of minced meat – usually beef – with breadcrumbs, garlic and herbs, are fried in olive oil and sometimes served with a tomato sauce. *Keftédes* is a very popular dish in the islands, especially on Crete, and is found on most restaurant menus.

Beef *sofríto*, a traditional stew

9 Sofríto

Although most popular on Corfu, where it originated, this dish is now increasingly seen elsewhere on the islands. It is made using a delicious recipe featuring veal cooked slowly in a creamy white wine, garlic and herb sauce, traditionally topped with herbs and served with rice and salad.

10 Saganáki

This dish, which is normally eaten for lunch, consists of cheese cooked in a frying pan *(saganáki)* and served with bread. The ingredients added to the pan can vary, but usually include seafood, onions or just lemon juice. The dish is quite often flambéed at the table.

TOP 10 MEZÉ DISHES

A dish of *tzatzíki*

1 Tzatzíki
A light dip that is always served cold, *tzatzíki* is produced using strained yoghurt, garlic and cucumber.

2 Taramosaláta
This dip is made from *tarama,* the salted and cured roe of mullet or cod, mixed with olive oil and lemon. It is often served as a starter with bread.

3 Tahíni
Made with sesame seeds pureed with lemon and garlic, this *mezé* or appetiser dip has a distinctive flavour.

4 Dolmádes
Vine leaves stuffed with currants, rice and pine nuts, *dolmádes* is a popular *mezé* dish seen on many menus.

5 Skordaliá
A dip or thick sauce originating from the Ionians, *skordaliá* is made of mashed potato, garlic and lemon.

6 Loukánika
Loukánika is pork sausage flavoured with fennel and orange, and it is usually served sliced as a *mezé* dish.

7 Olives with Garlic
Black and green olives are always served as a part of *mezédes*; garlic is often added, giving them bite.

8 Melitzanosaláta
Aubergines are cooked and pureed with tomatoes, garlic and lemon to create this delicious dip.

9 Revithosaláta
Sometimes known as hummus, *revithosaláta* is a Greek dip made with chickpeas, garlic and coriander.

10 Choriátiki Saláta (Greek Salad)
This appetizing blend of tomatoes, cucumber, peppers and olives is topped with feta cheese and basil.

🔟 Restaurants

The pretty terrace at Tassia, an ideal place for alfresco dining

1 Tassia, Kefaloniá

Run by cookbook author Tassía Dendrínou, who still does much of the cooking here herself, this popular restaurant is housed in a delightful Venetian building that overlooks Fiskárdo's harbour. The menu features dishes from Tassía's own books, including traditional Kefallonian meat pie and lobster spaghetti. The extensive wine list has international and local vintages *(see p87)*.

2 Taverna Karbouris, Corfu

One of Corfu's oldest north-coast tavernas, Karbouris has become a local landmark. Traditional Greek cuisine, such as *kléftiko*, is served at this family-run establishment. You can sit near the pool or inside in the attractive dining room *(see p83)*.

3 Remetzo, Égina

A harbourside fish restaurant in delightful Pérdika. The menu offers starters like Greek salad and dips, alongside an excellent choice of seafood dishes that include squid cooked with tomatoes and herbs and delicious prawns with *ouzo*. Join locals to watch the sun go down *(see p155)*.

4 La Maison de Catherine, Mýkonos

This classy restaurant is known for its artfully presented tables, à la carte menu of Greek dishes with a French twist and its fine wine list. Classical music playing in the background enhances the experience. Be sure to try the prawns in a delicious tomato and *oúzo* sauce. Elegant touches like crisp linen make this a memorable place *(see p95)*.

La Maison de Catherine's courtyard

Theodosi, Crete

One of the best restaurants in Chaniá, Theodosi has a refreshing approach to cuisine. Its imaginative Cretan menu uses only the freshest seafood from the Mediterranean and organic produce direct from the farmer. A house speciality is steamed mussels with saffron and cream. Fine wines and espresso complete the experience (see p108).

 Mesogeios Restaurant, Kárpathos

Located on the waterfront, this lively restaurant is known for its good selection of local dishes and its friendly staff. Diners sit at tables and chairs painted the colour of the sea, and can select from a menu that includes stuffed clams, squid or artichokes, goat *stifádo* (stew) and *makarónia*, a delicious local pasta. Right next to Mesogeios is its twin restaurant, Didyma, an integrated grill house (see p119).

 Alexis 4 Seasons, Rhodes

Housed in a period mansion with a roof garden and lounge bar, this award-winning restaurant features a gourmet-style menu with Mediterranean and Greek influences. The emphasis is on seafood and the vegetables come from its own garden. Specialities include smoked salmon, scallops in white vodka sauce and prawn risotto. A fine wine list, immaculate staff and elegant surroundings all complete the dining experience offered here (see p117).

 Sirines Restaurant, Itháki

Located in Ithaki Yacht Club, Sirines Restaurant is known both for its menu and the collection of shipping memorabilia that adorns the walls. Run by Marína Fotopoúlou, who specializes in Ithakian cuisine, it offers delicious dishes such as langoustine prawns in a mustard and lemon sauce and vegetarian *moussakás*. Organically grown vegetables are used (see p81).

 Melitini, Santoríni

This delightful tapas restaurant in the picturesque village of Oía serves traditional Greek cuisine made with locally sourced ingredients. Offerings include cured delicacies and cheeses, heartier fare, such as fava bean dishes, spicy beef sausage, or grilled octopus, and homemade orange cake with mastic ice cream for dessert. The rooftop terrace affords stunning views of the caldera on clear days and is the perfect place for lunch (see p99).

Dining on the sand, Sea Satin Market

Sea Satin Market, Mýkonos

This waterfront restaurant, which is located below the famous windmills offering views of Little Venice's neighbourhood, is one of the trendiest in town. Diners can select their fish fresh from the day's catch at the ice counter, watch it being cooked on charcoal grills and then enjoy the exquisite flavour, or if preferred select a starter and a platter from the menu (see p95).

🔟 Arts and Crafts

1 Lace and Embroidery
Lace and embroidered cushions, tablecloths and bread baskets are still handmade in many villages. Karyá in central Lefkáda is famous for its needlecraft, known as *karsanía*, and has a small museum dedicated to the craft.

2 Leatherware
Craftsmen can be seen in small workshops making leather sandals or the traditional knee-high boots. Both are made by hand as they have been for centuries. Handcrafted leather purses, handbags and belts are also popular souvenirs.

3 Icons
Religious icons depicting the Virgin and Child, and saints and angels, have been created for centuries, many originating on Crete. St Luke is credited with being the first iconographer and his icons are considered sacred in the Greek Orthodox Church. Visit churches to see genuine icons.

4 Carved Wooden Spoons
Normally made from olive wood, these spoons are the result of a craft handed down through generations. Wood is crafted into the shape of a spoon and its handle can be ornate or simply a figure.

Gift shop selling wooden spoons

Traditional Cretan blue pottery

5 Ceramics
Individual islands have their own style of painting ceramics. Égina is known for its more flowery designs, while Mýkonos artisans often paint traditional Greek figures. Villages are the best place to see authentic items and rural ceramics tend to be left unglazed rather than the colourful glazed items you will find in gift shops.

Copper coffee pot

6 Copperware
Today copperware tends to be used for decorative purposes only, but there was a time when local communities in Greece relied on the metal to make everyday household pots and pans. The traditional pot with a handle used to make Greek coffee, a *bríki*, is still in daily use to this day.

7 Jewellery

Gold and silver are a good purchase in the islands and main towns offer a wide choice of modern and traditional designs. Often the pieces echo ancient Greek styles. In Crete, for example, many items are styled on Minoan jewellery. Rustic handmade pieces may be found in the villages.

8 Rugs

The art of making a genuine sheep- or goat-wool rug, known as a *flokáti*, has a very long tradition in the Greek Islands. It dates from the 5th century, when mountain shepherds wore the rugs for warmth. Nowadays they are more likely seen covering the stone floors of village homes, and most rugs are made by machine.

Brightly coloured rugs on display

9 Beads

Known as *kombolói* (worry beads), these strings of beads can be found in gift shops, markets and jewellers throughout the islands. Greeks traditionally thread them through their fingers for relaxation. Some strings are coloured and worn as a fashion accessory.

10 Wall Hangings

Weaving thick cloth and mounting it on a top frame for hanging on a wall is an old craft still practised today in Greece. Locals will have several of these hangings in their home as decoration. They are commonly sold in gift shops and make excellent souvenirs.

TOP 10 POPULAR SOUVENIRS

Honey, an ideal gift to take home

1 Honey
Honey, sometimes flavoured with thyme, is produced and used in traditional cakes and desserts.

2 Olive Oil
Bottles of oil produced from the olives grown on the islands can be found in gift shops and supermarkets.

3 Herbs and Spices
Grown in quantity throughout the islands, herbs and spices can make good souvenirs to take home.

4 Leather Sandals
Leather sandals made in a village workshop are useful, as well as being reminders of the holiday.

5 Plaques
Sculpted in wood, stone and resin, plaques are usually fairly ornate and depict scenes of rural life.

6 Baklava Pastries
Made of *fýllo* pastry layers with nuts and drenched in honey, baklava suits a sweet tooth and makes a great gift.

7 Jewellery
Strings of coloured worry beads and items of gold or silver are popular items bought as souvenirs.

8 Ornate Pottery
Pottery jugs, bowls, candle holders and plates painted in traditional designs are ideal to take back home.

9 Glyká Koutalioú Confectionery
Candied pieces of fruit, nuts and sometimes vegetables, these sweets, full of calories, are mouthwatering *(see p62)*.

10 Embroidered Linen
Often trimmed with lace, these cushions and tablecloths provide a beautiful and elegant reminder of a holiday in the Greek Islands.

🔟 Greek Islands for Free

The stunning Navágio Beach on the island of Zákynthos

1 Beaches

Greece's islands have some of the finest beaches in the world, and most are totally free to visit. One of the most famous is Navágio Beach *(see p84)*, also known as Shipwreck Bay because of a freighter that sits in the sand. You'll see it on the cover of almost every guide to the Greek Islands. Be aware that on the larger more touristy beaches, such as those on Mýkonos, you might be charged for the use of a lounger.

2 Monasteries

No holiday in the Greek Islands would be complete without a visit to a monastery *(see pp44–5)*. The landscape is dotted with them, and most are free to visit on Sundays. If you're holidaying on Híos, head for the 11th-century Néa Moní *(see pp20–21)*, where you can see its famous mosaics (open until sunset).

3 Museums

The Greek Islands are home to some fabulous museums, like the Irákleio Archaeological Museum on Crete *(see pp27–8)* with its world-famous Minoan collection. Many have free admission days, such as the International Museums Day on 18 May, and most also have at least one Sunday a month when you can visit for free.

4 Festivals

The Greek calendar is bursting with annual events, such as Epiphany and Pentecost *(see p45)*, which are traditionally held in the harbours or on the seafronts of the islands. Whole communities come together to dance and play traditional music at these festivals, which tourists can attend for free. August is the best time of the year for village fetes, which are great events for mingling with the locals *(see pp70–71)*.

5 Sunsets

The islands are awash with quiet spots where visitors can enjoy a spectacular view of the setting sun. Falásarna Beach *(see p105)* on the western shores of Crete and the picturesque village of Pélekas *(see p53)* in the west of Corfu offer some of the best viewing opportunities.

Sunset over Chaniá lighthouse, Crete

6 Archaeological Sites

Many of the Greek Islands' archaeological sites are free on the first Sunday of the month (November–March). They include the Palace of Knossos *(see pp30–31)*, Delos *(see pp18–19)* and Heraion *(see pp24–5)*.

7 Walking Tours

Visit the tourist office in the main town of each island as most have a selection of free walking tours. On Skópelos, for example, a tour enables you to see many of the 100 or so churches that line the bay *(see p136)*.

8 Arts and Crafts

Galleries selling arts and crafts are dotted throughout the islands, all offering you the chance to take in displays and exhibitions for free. If visiting the increasingly trendy Sporádes *(see p143)* you'll be spoilt for choice.

A pretty street in Corfu Old Town

9 Explore the Old Quarters

You can take in the fascinating history of quarters like Corfu Old Town *(see pp12–13)* without spending a penny. Rhodes Old Town *(see pp14–15)*, the cobbled streets of Chaniá's trendy Spiántza Quarter *(see p102)* and Hydra Town *(see p147)* in the Argo-Saronics are just a few places where you can see some incredible sights for free.

10 Walking and Hiking Routes

Tourist offices have free guides to hiking routes of varying distances and terrains. Many take in sights too, for example, the Paleochora to Vathý route on Itháki *(see p75)* stops for a break at the Cave of Nymphs.

TOP 10 MONEY-SAVING TIPS

Greek salad, a typical local dish

1 When eating out, order local dishes and wine rather than expensive imports.

2 Visit the Greek Islands in late autumn or early spring to get the best deal and avoid the crowds. Enjoy the beautiful autumn colours or spring flowers and take walks in pleasant temperatures.

3 It can often be cheaper to book a low-cost flight and accommodation separately, rather than opting for a package deal. If you can be flexible on dates, you are likely to pay a lot less.

4 Students and senior citizens can get reduced admission at many museums and archaeological sites with identity cards and proof of status.

5 Shopping at local markets and staying in self-catering accommodation normally works out much cheaper than dining out in hotel restaurants.

6 Picnics are a favourite summer event for Greek families. Shop for fresh picnic ingredients and take them to the beach or a designated picnic spot.

7 The local tourist offices can provide a calendar of free events, which can range from village festivals to cultural events. Visitors are always welcomed by locals and encouraged to join in.

8 Rather than taking the relatively expensive taxis, travel by public bus – it's a great way to mingle with locals.

9 Hitchhiking is permitted on the islands and most Greeks tend to be generous in offering lifts, especially in remote areas, but always take sensible precautions to stay safe.

10 Camping can be cost-effective and fun. There are some great sites, usually well-equipped and close to the beach or in forests or olive groves.

🔟 Festivals and Events

1 Festivals for Patron Saints

Ágios Nikodímos Day on Náxos and the Litany of Ágios Spyrídon in Corfu are just two of the many festivals that take place around the Greek Islands, commemorating each island's patron saint. Festivals usually involve music, dance, parades, fireworks and feasts.

The Festival of St Paul, Rhodes

2 Hippocrátia Festival, Kos

The highlight of the annual Hippocrátia Festival is a recital of the Hippocratic Oath. Kos, Hippocrates' birthplace, celebrates with educational displays, folkloric exhibitions, music and theatrical shows. The festival runs from July to September.

3 Cultural Festivals

The islands are known for their festivals of art, literature, history and theatre. These include the Festival of the Aegean, held in Sýros each July, which involves opera and performances of the plays of Shakespeare. In May, Rhodes hosts its Medieval Rose Festival, while in August, Lefkáda hosts its Festival of Art and Literature.

4 Ochi Day

A public holiday on 28 October commemorates the day in 1940 when the Greek Prime Minister, Ioánnis Metaxás, rejected Mussolini's ultimatum to allow Axis forces to occupy parts of Greece. Metaxás exclaimed "ochi", meaning "no", and this one word marked the start of Greece's involvement in WWII.

5 Religious Festivals

After Easter and Christmas, the most celebrated religious day in Greece is the Dormition of the Virgin, or Feast of Assumption, on 15 August. It is believed that on this day the Virgin, Mary the Theotókos, ascended to heaven. Families celebrate with prayer, a feast and traditional dancing and music (see p45).

Traditional dancing performance, a big part of religious festivals in Greece

Festival in the square of Castro, Sikinos

6 Folklore Festivals

It is the tradition for every village in Greece to host a folklore festival *(panegýri)* at least once a year, usually during the summer. It is a time when people meet up with friends, feast together and dance. Musicians play traditional songs, while dance troupes perform folkloric dances.

7 International Festivals

Among the many festivals held on the islands are the Winter Festival in Kos, international music and dance, the Argostóli International Choral Festival in Kefalloniá (August) and the International Music Festival in Santoríni (September).

8 Name Days

Most children born in Greece are given the name of a saint. On that saint's feast day everyone with the same name celebrates. For example, 7 January is the name day for Ioánnis (John). In Chaniá on Crete, males with this name will visit the Church of Ágios Ioánnis and then party.

9 Town Carnivals

February is the carnival month and every town puts on events lasting about a week, usually culminating in a procession of floats. Most people, especially children, wear fancy dress.

10 Food and Wine Festivals

A successful food harvest or wine year is celebrated in style. On Kos there is a Fish Festival and a Honey Festival in August, while Crete has a Chestnut Festival in October.

TOP 10 FILMS SHOT IN THE GREEK ISLANDS

1 Mamma Mia, Skiáthos
A 2008 film starring Meryl Streep and Pierce Brosnan, which was based on the music of Swedish pop group ABBA.

2 Zorba the Greek, Crete
Based on the book by Cretan writer Nikos Kazantzákis, this 1964 film starred Anthony Quinn.

3 Tomb Raider: The Cradle of Life, Santoríni
A 2001 production starring Angelina Jolie as Lara Croft, this box-office hit film was shot on Santoríni.

4 For Your Eyes Only, Corfu
The casino scene of this 1981 film starring Roger Moore as 007 was shot in the Achílleion Palace, Corfu *(see p82)*.

5 Shirley Valentine, Mýkonos
This 1989 film starred Pauline Collins as a bored housewife on holiday in Greece finding herself.

6 Bourne Identity, Mýkonos
The concluding scenes of this 2002 film starring Matt Damon were shot here.

7 Guns of Navarone, Rhodes
Gregory Peck and David Niven were the big names in this 1961 film about German-occupied Greece.

8 Boy on a Dolphin, Hydra
Starring Italian actress Sophia Loren as sponge-diver Phaedra, this 1957 film was what made Hydra famous.

9 Big Blue, Amorgós
Released in 1988, the popular English-language version of the French classic *Le Grand Bleu* starred actor Jean Reno.

10 Captain Corelli's Mandolin, Kefalloniá
Nicolas Cage and Penelope Cruz were in this 2001 film that was set during the Italian occupation.

Captain Corelli's Mandolin, 2001

Greek Islands
Area by Area

Traditional architecture silhouetted
against a sunset in Oía, Santoríni

TOP 10 The Ionians

The Ionian islands sit in turquoise waters off the western coast of Greece and consist of six main islands: Kefalloniá, Corfu, Zákynthos, Paxí, Itháki and Lefkáda. There are also many smaller inhabited islands and tiny islets. The islands have been populated by ancient Greek, Roman, Byzantine, Venetian, French and British citizens, all of whom have left their legacy on this fabulous mixture of mountains, forests, beaches, bustling towns, idyllic villages and ancient sites.

Pórto Katsíki beach, one of Lefkáda's most beautiful spots

THE IONIANS

- ① **Top 10 Sights** see pp75–7
- ① **Restaurants: Paxí, Lefkáda, Itháki** see p81
- ① **Bars and Cafés** see p80
- ① **Best Beaches** see p78
- ① **Islands and Islets** see p79

See Corfu map

Kónitsa
Ioannina
Glyki
Paxí
Andípaxi
Árta

0 km 40
0 miles 40

② **Lefkáda**
Lefkáda Town
Nydri
Nydri Waterfalls

0 km 10
0 miles 10

See Lefkáda map

See Itháki and Kefalloniá map

See Zákynthos map

Corfu

Períthia
Sidári
Róda
Achavári
Kassic
Ágios Stefanos
Afionas
Corfu
Kalám
Skriperó
Barbatí
Dasía
Paleokastrítsa
Gouviá
Kontokáli
Cor
Ermones
Tov
Pélekas
Kalafatiónes
Ágios Górdios
Beni
Ágios Ioánnis
Ágios Matthaíos
Mesong
Chalikoúnas
Argyráde

Ionian Sea

- ① **Places to Eat: Corfu** see p83
- ① **Sights: Corfu** see p82

0 km 6
0 miles

1 Paxí
MAP B2

Covered in olive groves, Paxí is a pretty island with a busy capital, Gäios. Two islets in the bay, the fortified Ágios Nikólaos and Panagía, provide a natural breakwater. The rugged coastline is interspersed with small coves and fishing villages like Longós and Lákka. Magaziá is famous for its sunsets.

2 Lefkáda
MAP G1

Archaeological Museum: Ángelou Sikelianoú; Tel 26450 21635; Open 8am–3pm Tue–Sun; adm

This mountainous island is densely forested. Connected to the mainland by bridges, its capital, Lefkáda Town, was rebuilt using brightly coloured metals after a 1948 earthquake. Don't miss the Archaeological Museum *(see p42)*. Nydrí and Vasilikí, along with Pórto Katsíki beach *(see p50)*, are popular, while offshore lie the Meganíssi, Madourí and Spárti islands *(see p79)*.

Statue of King Odysseus, Itháki

3 Itháki
MAP H2

Famous for being mentioned in the works of Homer, Itháki is thought to be the home of the mythological King Odysseus. It has mountains and cliffs to the west and dense forest to the east. The island's capital is the charming Vathý.

Ithaki ③ and Kefalloniá ⑨

① Places to Eat: Kefalloniá see p87

① Sights: Kefalloniá see p86

Zákynthos ⑩

① Places to Eat: Zákynthos see p85

① Sights: Zákynthos see p84

GERALD DURRELL

When Durrell moved to Corfu in 1935, the island's flora and fauna developed his love of animals. Durrell's Corfiot years were recounted in his *Corfu Trilogy*, comprising *My Family and Other Animals* (1956), *Birds, Beasts, and Relatives* (1969) and *The Garden of the Gods* (1978).

Andípaxi
MAP B2

Andípaxi has a population of less than 60 permanent inhabitants. It is a haven of clear turquoise waters and golden beaches, including Vríka beach, nestled in a sheltered cove, and nearby Voutoúmi beach. It is possible to take a walk along the coastal path that links the two. Andípaxi, which can be reached by boat from Gäios on Paxí, is covered with vineyards, which provide the grapes used to make its local wine.

Caves of Sámi, Kefalloniá
MAP G3 ■ Drogaráti and Melissáni caves: open 8:30am–8pm daily; adm

One of Sámi's many caves, the huge Drogaráti cave is believed to be more than a million years old and has hundreds of stalactites and stalagmites. With superb acoustics, the creatively lit cave is often used for cultural and musical events. Nearby is the Melissáni cave, said to have been the Mycenaean sanctuary of the god Pan. Its lake is deep blue, highlighted by the sun shining through a hole in the cave's roof. There are other minor caves too.

Corfu
MAP B2

Miles of unspoiled coastline, dotted with resorts, inland villages where little has changed in centuries and forested countryside, make up Corfu. The Paleokastrítsa headland is one of its natural wonders, as is the wildlife haven of Korissíon Lagoon and the island's highest point Mount Pantokrátor, while Sidári and Benítses are bustling holiday spots *(see p82)*. It's capital is also named Corfu and the Old Town boasts many historic sights *(see pp12–13)*.

Mitropoli Panagias, Corfu Old Town

Mýrtos Bay, Kefalloniá
MAP G2 ■ Near Divaráta

Considered the most photographed bay on the island, complete with a stunning beach *(see p78)*, Mýrtos Bay lies between Mount Kalón Oréon and Mount Agía Dynatí, famous in Greek mythology as the rock thrown by Cronus, leader of the Titans.

8 Nydrí Waterfalls, Lefkáda
MAP G1

Lefkáda has a rare beauty, and nature has created some spectacular gorges here, with sheer rock faces and waterfalls. The Nydrí Waterfalls, which lie to one end of the Dimosári gorge near Perigiáli and Nydrí, are the most impressive. Underground caverns of water fuel the cascades, which plunge at an astonishing rate to crystal-clear pools. Visitors can take a refreshing dip in the wondrous waterfalls' pools (see p54).

9 Kefalloniá
MAP G2

The largest Ionian island, Kefalloniá is green and mountainous. It is best known for its vineyards, subterranean waterways, caves and long, indented coastline. Mount Énos (Aínos) is over 1,600 m (5,200 ft) high and carpeted with black fir trees and old olive trees. Kefalloniá's main towns and villages include Sámi, the capital Argostóli, Fiskárdo, Skála and Lixoúri (see p86).

10 Zákynthos
MAP B4

Zákynthos is a gem of an island with high mountains clothed in cypress trees and some stunning beaches, the most famous of which are Navágio and Laganás in the south (see p78). Its bustling capital, Zákynthos Town, was rebuilt to its original Venetian look and layout after being destroyed in the 1953 earthquake. Places to visit include the beautiful Blue Caves (see p84).

Zákynthos Town, seen from the sea

DAY TRIP FROM CORFU TO PAXÍ

MORNING

The Ionian islands are scattered over a large area and not given to island hopping in the same way as the other Greek island groups. However, it is still possible to get from one island to another relatively easily. Corfu, Kefalloniá and Zákynthos, for example, all have airports that offer domestic flights, while yacht charter companies offer boat hire. Ferries also ply between most of the islands.

One of the most popular ferry routes is a day-trip to Paxí (see p75) from Corfu. Start your day at the New Port in **Corfu Town** and look for the signs indicating where you can catch the Flying Dolphin hydrofoil. There are daily departures in summer, but the first departure time varies depending on the day of the week. On board you can relax before disembarking at **Gáios New Port** on Paxí. The journey time is about an hour. In **Gáios** you can hire a car or taxi to take you around the island. Head to the fishing village of **Longós** for a delicious lunch at one of the waterfront tavernas.

AFTERNOON

After a leisurely lunch take the road signposted to **Lákka**. Be sure to explore the town before driving south past the communities of **Magaziá** and **Oziás**, then back to Gáios. Take some time to investigate the harbourside and enjoy a meal before you leave, and then hop on your hydrofoil, which departs at around 7pm, for the return journey.

See map on pp74–5

Best Beaches

Myrtiótissa Beach, sheltered by cliffs

1 Myrtiótissa Beach, Corfu
MAP A5 ■ West coast

Sheer cliffs and smooth golden sand characterize this horseshoe-shaped beach, which has been described as Europe's most beautiful. It is favoured by naturists because of its seclusion.

2 Laganás Beach, Zákynthos
MAP H4

This beach is particularly sandy, making it popular not only with visitors, but also with the endangered loggerhead turtles, the *caretta caretta*, which lay their eggs here. It is great for snorkelling (see p84).

3 Megálo Limonári Beach, Meganíssi
MAP H1 ■ Near Katoméri

Surrounded by forest and with beautiful soft sand, this isolated beach is tucked in a small bay on Meganíssi island. To find it, look out for signposts from Katoméri village (see p50).

4 Arkoudáki Beach, Paxí
MAP A2 ■ Near Lákka

This secluded sandy beach can only be reached by boat and is therefore quiet. It is a favourite spot with people coming ashore from their yachts anchored in the bay.

5 Paleokastrítsa Beach, Corfu
MAP A5 ■ Paleokastrítsa

A favourite of Sir Frederick Adam (see p82), this sandy beach is quiet, with just a handful of nearby tavernas. Offshore lies the Liapádes reef, a popular diving and snorkelling spot.

6 Kalamítsi Beach, Lefkáda
MAP G1 ■ Kalamítsi

There are interesting caves and rocks for snorkellers to explore around this west coast beach. For those who like to take it easy, the sand and shingle beach is soft.

7 Pórto Katsíki Beach, Lefkáda
MAP G1 ■ Near Vasilikí

With its horseshoe-shaped bay and high cliffs, forested backdrop and curious offshore rock formations, this is among the most picturesque beaches on Lefkáda (see p49).

8 Gidáki Beach, Itháki
MAP H2 ■ Near Vathý

With a hiking track as its only access, Gidáki beach is unknown to most visitors and is a great place to relax. Its bay, however, attracts some boats from Vathý during summer.

9 Navágio Beach, Zákynthos

This white sandy beach is protected by vertiginous limestone cliffs on both sides. The beach is only accessible by boat. There are frequent services from the small port of Porto Vromi or from Zákynthos Town (see p84).

10 Mýrtos Beach, Kefalloniá
MAP G2

Mýrtos Bay is famous for its dazzling turquoise waters and a beach of white pebbles that decrease in size until they shelve deeply into the sea, a process known as longshore drift. The quiet and unspoiled beach lies between two mountains (see p86).

See map on pp74–5

Islands and Islets

1 **Ereíkoussa Island**
MAP A1 ■ Corfu

Characterized by its cypress trees, great beaches and the small town of Pórto, where most of its inhabitants live, Ereíkoussa is one of the Diapontian islands and can be reached by boat from Sidári.

2 **Othonoí Island**
MAP A1 ■ Corfu

According to Greek mythology, the island of Othonoí is where Odysseus met Nausika. Offering a peaceful way of life, it is the largest Diapontian island, located between Mathráki and Ereíkoussa.

The house of Aristotélis Valaorítis on Madourí Island

3 **Madourí Island**
MAP H1 ■ Lefkáda

In the bay off Nydrí, Madourí is a heavily forested island. Its sheltered coastline makes it a popular spot to sail around. The Greek poet Aristotélis Valaorítis (1824–1879) once lived here.

4 **Mathráki Island**
MAP A1 ■ Corfu

This quiet island manages to avoid tourists except occasional hikers that brave the rocky coastline. The reefs, a rich seabed and beaches with crystal clear water make it a scenic attraction.

5 **Spárti Island**
MAP H1 ■ Lefkáda

Spárti is covered with dense forest. Boats on day trips drop anchor in the shallow water and sheltered coves of its slightly indented coastline.

6 **Diá Island**
MAP G3 ■ Off Argostóli, Kefalloniá

Sometimes known as Theionísi, Diá island is the source of many legends. The most popular of these is that it was home to a temple dedicated to the Greek god Zeus.

7 **Skorpídi Islet**
MAP H1 ■ Lefkáda

A small islet off the private Skorpiós island, Skorpídi is known for its wildlife. You can take a boat from Nydrí and anchor up for a lovely day out.

8 **Lazaréto Islet**
MAP H2 ■ Itháki

When passing by in the harbour of Vathý, the serene islet of Lazaréto is a sight to behold with its pretty white-washed chapel among dense trees.

9 **Strofádes Islands**
MAP B4 ■ Near Zákynthos

Part of the Zákynthos National Sea Park, the Strofádes islands of Arpiá and Stamfáni (plus a few smaller islets) are uninhabited except for a lone monk who is said to live there.

10 **Meganíssi Island**
MAP H1 ■ Lefkáda

Rural Meganíssi is the largest island off Lefkáda. It has quiet beaches and just three villages, including pretty Katoméri and the capital Vathý, which has a harbour lined with yachts.

Fisherman mending his net, Vathý

Bars and Cafés

1 Palazzo Cocktail Bar, Corfu

MAP A4 ▪ Sidári ▪ 26630 95946

As well as offering a range of draught beers and ciders, cocktails, wine and non-alcoholic drinks, Palazzo also serves food all day. A traditional English breakfast is followed by international meals *(see p61)*.

2 Genesis Taverna and Café Bar, Paxí

MAP B2 ▪ Gäios ▪ 26620 32495

Renowned for its doughnuts, this café also serves a wide range of drinks, snacks and ice creams, and has views across Gäios Harbour.

3 Bass Club, Kefalloniá

MAP G3 ▪ Argostóli ▪ 26710 25020

One of the hottest enter-
tainment venues in Kefalloniá, this club plays non-stop dance music, with both Greek and international hits. There are also regular performances from guest DJs and live bands *(see p60)*.

4 DiZi Bar, Corfu

MAP A5 ▪ Ermones ▪ 26610 14069

This 24-hour bar serves fine coffee during the day and offers free Wi-Fi. Guests can enjoy superb cocktails and live music after dark *(see p60)*.

Outside seating by the water, Corfu

5 Café Bar Muses, Kefalloniá

MAP G3 ▪ Lourdáta ▪ 26710 31175

A lively café-bar, Muses has a menu of light snacks and exotic-looking cocktails, a lounge with screens showing live sports and a garden terrace with a children's play area.

6 Karamela Café, Itháki

MAP H2 ▪ Harbour, Vathý ▪ 26740 33580

In a brightly painted building on the harbourside, this café offers an assortment of honey-drenched pastries, cakes and beverages. Enjoy evening cocktails and live music beside the water.

Ice cream at Genesis

7 Rescue Club, Zákynthos

MAP H4 ▪ Laganas ▪ 26950 51612

Rescue Club has gradually risen in popularity ever since it was established in 1987, and can now call itself the top club in Zante *(see p61)*.

8 Saratseno Café Bar, Zákynthos

MAP H4 ▪ Tsiliví ▪ 26950 45738

This lively café-bar has widescreen TVs, Greek-themed evenings and bingo and table games nights.

9 155 Cocktail Bar, Lefkáda

MAP G1 ▪ Vasilikí ▪ 26450 31868

This venue's chic decor, central location, 60-plus cocktails and non-alcoholic "mocktails", and jazz tunes playing in the background, make it a fashionable place to relax.

10 Sea to See, Lefkáda

MAP H1 ▪ Angelou Sikelianou, Lefkáda Town ▪ 26450 25365

On the seafront, this dynamic eatery serves international fare. Tuck into steak, grilled chicken, and breakfasts, and finish with cold frappés.

Restaurants: Paxí, Lefkáda, Itháki

PRICE CATEGORIES

For a three-course meal for one with half a bottle of wine (or equivalent meal), taxes and extra charges.

€ under €30 €€ €30–€50 €€€ over €50

1 Nassos, Paxí
MAP B2 ▪ Longós harbour ▪ 26620 31604 ▪ €€€

Great for lunch or dinner, Nassos is known for its local fish such as sword-fish, sea bream, octopus and squid, which are served in tangy sauces.

2 Zolithros, Lefkáda
MAP H1 ▪ Mikros Gïalos ▪ 69723 18385 ▪ €

Serving fish straight from the owner's own boat, Zolithros is a classic blue and white taverna on the harbourside.

3 Poseidon Fish Tavern, Itháki
MAP H2 ▪ Vathý ▪ 69737 80189 ▪ €€

Octopus, lobster and swordfish, all artfully plated, are specialities at this restaurant just off the harbour.

4 Trehantiri, Itháki
MAP H2 ▪ Vathý ▪ 26740 33444 ▪ €

Locals as well as visitors beat a path to Trehantiri for delicious homemade dishes using family recipes.

5 Taverna Vassilis, Paxí
MAP B2 ▪ Main Square, Gäios ▪ 26620 31587 ▪ €€

This traditional eatery has a wide choice of classic Greek dishes. Try the *moussakás* or something different, like octopus in red wine.

6 Sirines Restaurant, Itháki
MAP H2 ▪ Ithaki Yacht Club, Vathý ▪ 26740 33001 ▪ €€

This restaurant is popular for its international fare, as well as local specialities such as *Strapatsada*, clay pot lamb, rooster and more *(see p65)*.

7 Taka Taka, Paxí
MAP B2 ▪ Gäios ▪ 26620 32329 ▪ €€

Opened in 1970, this restaurant has a faithful clientele and is known for its delicious char-grilled fish dishes served with herbs and lemon. Dine inside or on its vine-covered terrace.

8 Basilico, Lefkáda
MAP H1 ▪ Harbour, Nydrí ▪ 26450 61632 ▪ €€

This popular eatery on the waterfront is run by a local family and serves classic Greek dishes with great flair.

Inviting terrace at Basilico

9 Rachi, Lefkáda
MAP G1 ▪ Exanthia ▪ 26450 99439 ▪ €

Enjoy panoramic views of the sea from the spacious veranda at this modern waterfront restaurant, which offers a lengthy selection of classics. Take in the sunset with a cocktail.

10 Fryni Sto Molo, Lefkáda
MAP H1 ▪ Golémi, Lefkáda Town ▪ 26450 24879 ▪ €

Fryni Sto Molo is known for its *mezédes*, a traditional way of eating where small portions of local dishes are brought to your table at intervals. Beer is served cold from the cask.

See map on pp74–5 ←

Sights: Corfu

Canal d'Amour rock formation, Sidári

1 Sidári
MAP A4

The Pre-Neolithic remains found here reveal Sidári's long history. It is now a tourist hotspot that is famous for its rock formations, which have created a channel known as the Canal d'Amour.

2 Achílleion Palace
MAP B5 ▪ Gastoúri, Corfu Town ▪ 26610 56245 ▪ Open 8am–7pm Mon–Fri, 8am–2:30pm Sat & Sun ▪ Adm

Empress Elisabeth of Austria, known as Sissi, created this Neo-Classical palace in the 1890s. Statues of Greek gods adorn the house and gardens.

3 Mount Pantokrátor
MAP B4

At 900 m (3,000 ft), this is Corfu's highest mountain, which offers great views. Its tiny villages are set among olive groves and dense pine forests.

4 Korissíon Lagoon
MAP B6

This 5-km- (3-mile-) long fresh-water lagoon is home to turtles and tortoises *(see p54)*.

5 Lefkimmi Town
MAP B6

The tall houses and tiny streets in this attractive riverside town, famous for its excellent wine, are good examples of local architecture.

6 Mon Repos Estate
MAP B5 ▪ Corfu Town ▪ 26610 41369 ▪ Open 8:30am–3pm Tue–Sun ▪ Adm

This grand estate was built in the 1820s by Sir Frederick Adam, second High Commissioner of the Ionians.

7 Kassiópi
MAP B4

Kassiópi offers a wide choice of harbourside tavernas. Important artifacts excavated at its Angevin castle ruins are displayed in the Archaeological Museum.

8 Gardiki Castle
MAP B6

The ruins of this bastion, built by a 13th-century duke, Michail Angelos Komninós II, are well preserved.

9 Archaeological Museum
MAP B5 ▪ Vrailá Arméni Street, Corfu Town ▪ 26610 30680 ▪ Closed for renovation ▪ Adm

The 590-BC Gorgon Medusa, a frieze from the Artemis Temple, is exhibited here.

10 Benítses
MAP B5

This is one of the liveliest tourist resorts on the island, with tavernas, children's play areas, watersports and a buzzing nightlife.

The Runner, Achílleion Palace

Places to Eat: Corfu

① Agnes Restaurant
MAP A5 ▪ Pélekas ▪ 26610 94997 ▪ €€

This attractive restaurant is known for its authentic Corfiot dishes prepared by Agnes, the owner, using organic ingredients. The menu includes *moussakás* and squid served with herbs.

② Trilogia Restaurant
MAP B4 ▪ Kassiópi ▪ 26630 81589 ▪ €€

An imaginative menu of appetizers, pasta, meat and seafood dishes, plus fine wines and a great location overlooking the sea combine to give this place an edge. It offers a great children's menu too.

③ The Venetian Well
MAP B5 ▪ Lili Desila 1, Corfu Town ▪ 26615 50955 ▪ €€

With its elegant, cottage-chic decor, this restaurant is the place to eat if you want to be impressed by artfully presented dishes and fine wines. Dine inside or out in the courtyard.

④ Vergina
MAP A5 ▪ Gouviá ▪ 26610 90093 ▪ €

Traditional Greek music, dancing and Corfiot dishes are the order of the day at this taverna, housed in what was once a village bakery.

⑤ Anthos Restaurant
MAP B5 ▪ Maniarizi Arlioti, Corfu Town ▪ 26610 32522 ▪ €€

Inside a historic Corfiot town house, this trendy place is popular with locals. Dine on Greek dishes with inspiration from French bistrostyle cuisine, with wines to enhance.

⑥ La Famiglia
MAP B5 ▪ Maniarízi Antioti 16, Corfu Town ▪ 26610 30270 ▪ €€

Resembling an Italian trattoria and serving authentic dishes and wines from Italy, La Famiglia is one of the best restaurants in the town centre.

PRICE CATEGORIES
For a three-course meal for one with half a bottle of wine (or equivalent meal), taxes and extra charges.

€ under €30 €€ €30–€50 €€€ over €50

⑦ Panorama Restaurant
MAP A5 ▪ Ágios Georgios Bay, Afionas ▪ 26630 51846 ▪ €€

Perched on a cliff overlooking the beach, this is the place to dine on local dishes while watching the sun set over the waters below.

⑧ Taverna Sebastian
MAP A5 ▪ Ágios Górdios, Sinarádes ▪ 26610 53256 ▪ €€

This taverna serves its own Corfiot recipes with a modern twist and accompanied by good house wine. Try the prawn *saganáki*.

Outside terrace, Taverna Karbouris

⑨ Taverna Karbouris
MAP B4 ▪ Ágios Spyrídon, Períthia ▪ 26630 98032 ▪ Closed Mon–Fri in winter ▪ €€

Classic Greek dishes like *pastitsáda* (beef in tomato) and *kléftiko* (lamb with herbs) are served in this stone taverna in a village setting. Eat inside or alfresco on the terrace (see p64).

⑩ Bistro Boileau
MAP A5 ▪ Kontókali ▪ 26610 90069 ▪ €

Artfully presented Corfiot dishes are served on the terrace at this stylish bistro. Special diets are catered for.

See map on pp74–5 ←

Sights: Zákynthos

A rare loggerhead turtle

1 Laganás Bay
MAP H4

Laganás Bay is a breeding ground for the endangered loggerhead sea turtle. These beautiful creatures come ashore to lay their eggs on the protected Laganás beach *(see p78)*.

2 Maherádo
MAP H4

Visit the 14th-century Agía Mávra church here, with its frescoes and well-preserved icons, one of which is believed to be miraculous.

3 Zákynthos Town
MAP H4

The island's capital and port, Zákynthos Town, is characterized by Neo-Classical architecture, elegant squares and a bustling harbourside. Sights include the Ágios Dionýsios church.

4 Museum of Solomós, Zákynthos
MAP H4 ▪ Plateia Agiou Markou, Zákynthos Town ▪ 26950 42714 ▪ Open 8am–3pm Tue–Sun ▪ Adm

Founded in 1959, this museum houses the relics of two famous poets from Zákynthos, Dionysios Solomos and Andreas Kalvos *(see p42)*.

5 Kerí
MAP H4

One of the few villages to escape destruction in the 1953 earthquake, Kerí retains original stone houses. Surrounded by vineyards, it produces some of the island's best wine.

6 Volímes
MAP G4

This mountain village, located in the northwest of Zákynthos, is best known for its traditional textiles and its Venetian Baroque church, Agía Paraskeví, which has the finest gilded iconostasis screens on the island.

7 Navágio Beach
MAP G4 ▪ Near Volímes

This spectacular cove, with its sheer cliffs and pristine blue waters, is one of the most photographed beaches in Greece. A viewing platform on the edge of the cliff offers great photo opportunities *(see pp48 and 78)*.

8 Anafonítria
MAP G4 ▪ Near Volímes

Zákynthos' patron saint, St Dionýsios, was a monk at the 14th-century monastery of Panagía Anafonítria, in this sleepy mountain village located on the northwestern part of the island.

9 Melinádo
MAP H4 ▪ Near Maherádo

This small, traditional village is best known for the temple of Artemis, where remains, such as coins and ceramics, were discovered.

10 Blue Caves
MAP H3 ▪ Cape Skinári

These caves have amazing sculpted rock formations and caverns, surrounded by a sapphire blue sea. Take a boat out for the best view *(see p54)*.

The Blue Caves, a natural wonder

Places to Eat: Zákynthos

① Lofos Restaurant
MAP H4 ■ Méso Gerakári ■ 26950 62643 ■ €

With panoramic views from its hilltop location, Lofos serves authentic Greek dishes using local produce. Dine on the terrace if the weather is good.

② Dennis Taverna
MAP H4 ■ Lithákia ■ 26950 51387 ■ €

A family-run taverna with nostalgic village memorabilia, this taverna is known for its delicious meat and fish dishes served char-grilled with herbs, salad and homemade olive bread.

③ Panos
MAP H4 ■ Kalamki Road, Laganás ■ 26950 52685 ■ €€

Diners eat on the pretty terrace of this trendy town-centre grill house. It serves a selection of Zákynthian dishes, including *kleftiko* that is flambéed at the table. All dishes are made with fresh ingredients sourced from the local market.

④ Prosilio
MAP H4 ■ Panton and Latta 15, Zákynthos Town ■ 26950 22040 ■ €€

Classic Zákynthian dishes are given a creative twist at this rather trendy upmarket eaterie. Enjoy excellent food and wine while watching the open kitchen's drama unfold.

⑤ Romios
MAP H4 ■ Tsiliví ■ 26950 22600 ■ €€

Serving Italian pizza with a choice of toppings, plus homemade pasta, Mexican nachos and Greek *mezédes*, this popular, family-orientated restaurant in the centre of Tsiliví caters for almost every taste.

⑥ Massa Restaurant & Bar
MAP H4 ■ Kalamaki ■ 26950 33279 ■ €€€

Creative Mediterranean fare is on the menu at this restaurant at Hotel Venus. Risotto, pasta, grilled meats and fresh seafood are accompanied by Greek wines, beers and cocktails. Choose to sit on the outdoor terrace.

⑦ Zakanthi Restaurant
MAP H4 ■ Kalamáki ■ 26950 43586 ■ €

Known for its homemade dips, such as *tzatzíki* made from local yoghurt, cucumber and garlic, and its classic dishes, this restaurant offers the option to dine in its garden.

The bar at Zakanthi Restaurant

⑧ Flocas Café
MAP H4 ■ Argássi ■ 26950 24848 ■ €

If you fancy mingling with locals while enjoying an informal meal of crêpes and pancakes (both sweet and savoury) or a Greek evening meal, then this is the place to come.

⑨ Buon Amici
MAP H4 ■ Kalamáki ■ 26950 22915 ■ €€

This Italian eatery serves pasta, oven-baked pizza, seafood dishes, and delicious homemade sauces.

⑩ Peppermint
MAP H4 ■ Argássi ■ 26950 22675 ■ €€

This restaurant grows its own produce in an organic garden. The menu includes special fish dishes and traditional Greek plates.

See map on pp74–5 ⟵

Sights: Kefalloniá

1 Mýrtos Bay
MAP G2 ▪ Near Divaráta

Widely considered one of the most beautiful bays in the world, Mýrtos Bay boasts exquisite turquoise sea and dazzling white sand *(see p76)*.

Mosaics in the Roman villa, Skála

2 Skála
MAP H3

The remains of Old Skála, which was destroyed in the 1953 earthquake, and a Roman villa with some well-preserved mosaics are worth seeing in this attractive beach-side resort.

3 Lixoúri
MAP G3

The island's second-largest community, Lixoúri is an elegant town on the coast of the Pallíki peninsula. It has been a popular tourist spot since the 19th century.

4 Argostóli
MAP G3

Kefalloniá's capital was almost destroyed in the 1953 earthquake. It was rebuilt partly in its original Venetian style, and life revolves around its main square, Plateía Valliánou.

The charming port, Fiskárdo

5 Lake Ávythos
MAP H3 ▪ Near Ágios Nikólaos

According to folklore, Lake Ávythos is bottomless. It is a habitat for diverse flora and fauna.

6 Sámi
MAP G2

With its traditional houses and lively harbour full of tavernas, Sámi is popular with visitors. It is famous for Melissáni lake and Drogaráti cave.

7 Mazarakata Mycenaean Cemetery
MAP G3

This is one of the oldest sites on the island. Excavations have revealed several 1600 BC Mycenaean tombs.

8 Ássos
MAP G2

One of the island's prettiest coastal villages, Ássos has traditional architecture and sandy beaches. A small isthmus joins it to a Venetian castle's remains atop an islet.

9 Lourdáta
MAP H3

Overlooking the bay, this picturesque fishing harbour lined with tavernas has clear water and a long sandy beach. Nearby are the remains of a small 13th-century monastery.

10 Fiskárdo
MAP G2

Pastel-coloured Venetian buildings and cypress trees characterize Fiskárdo. It is also known for its fish restaurants.

Places to Eat: Kefalloniá

PRICE CATEGORIES
For a three-course meal for one with half a bottle of wine (or equivalent meal), taxes and extra charges.

€ under €30 €€ €30–€50 €€€ over €50

1 Elli's
MAP G2 ■ Harbourside, Fiskárdo ■ 26740 41127 ■ €
Housed in a birghtly painted house beside the water, this elegant eatery specializes in fresh fish, but there are meat and vegetarian options too.

2 Siroco
MAP H3 ■ Main St, Skála ■ 26710 83613 ■ €€
Located in the centre of the village of Skála, Siroco serves a wide selection of fresh local dishes.

3 To Perasma
MAP G2 ■ Sea Front, Agía Efimia ■ 26740 61990 ■ €€
Overlooking the harbour, this friendly taverna is run by two brothers. The extensive menu offers regional cuisine and includes plenty of dishes made with locally caught fish.

4 Apostolis Tavern
MAP H3 ■ Skála ■ 26710 83119 ■ €
Known for its wholesome country dishes like veal *stifado* (stew) and lamb *souvláki* (kebab), the Apostolis offers a true taste of Kefalloniá. Produce and wines are sourced locally. Dine inside or alfresco on its terrace.

Lamb *souvláki*, Apostolis Tavern

5 Aquarius Restaurant
MAP H3 ■ Skála ■ 26710 83612 ■ €€
Salkitier, a dish of pork in a spicy mustard sauce, and *stifádo* cooked with veal are just two of the recipes you might find on the menu at this stylish restaurant.

Lorraine's Magic Hill

6 Lorraine's Magic Hill
MAP G3 ■ Lourdáta Beach ■ 26710 31605 ■ €
This family-run restaurant offers sweeping views from the terrace. All vegetables used are organic. Try the *moussaká*.

7 Mikelatos Family Restaurant 'Scandinavia'
MAP H3 ■ Skála ■ 26710 83240 ■ €€
Come here for authentic Greek food. After your meal, order some cocktails and dance the night away.

8 Tzivras
MAP G3 ■ Vasíli Vandórou 1, Argostóli ■ 26710 24259 ■ €
Known for authentic country dishes, such as *moussakás* and meat pie, Tzivras has offered a taste of rural Kefalloniá since 1933.

9 Casa Grec
MAP G3 ■ Stavrou Metaxa 12, Argostóli ■ 26710 24091 ■ €€
The menu here offers a good mix of Mediterranean styles, such as Greek, French and Italian, often combining them in one dish. Try the prawns.

10 Tassia
MAP G2 ■ Harbourside Fiskárdo ■ 26740 41205 ■ €€
Beside the bay, this restaurant serves dishes taken from the cookbooks by its owner, Tassía Dendrínou (see p64).

See map on pp74–5

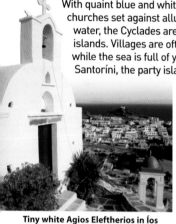 The Cyclades

With quaint blue and white houses, windmills and domed churches set against alluring beaches and crystal-clear water, the Cyclades are among the most stunning Greek islands. Villages are often perched on mountainsides, while the sea is full of yachts and fishing boats. Volcanic Santoríni, the party island of Mýkonos and the group's largest, Náxos, are cosmopolitan holiday destinations, while other islands are more traditional.

Tiny white Agios Eleftherios in Íos

1 Íos and Folégandros
MAP E5

In antiquity Íos is famous for being the burial place of the ancient Greek poet Homer, and each year a festival, the Omíra, is held in his honour. This mountainous island is speckled with chapels and tiny villages, while its

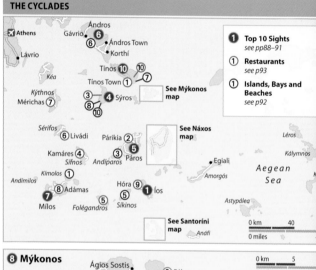

THE CYCLADES

Athens · Lávrio · Ándros · Gávrio **6** · **6** Ándros Town · Korthí · Kéa · Tínos **10** **10** · Tínos Town **1** **7** · Kýthnos · Mérichas **7** · Sérifos · **6** Livádi · **3** **4** Sýros · **8** **10** · Páraki **2** · Kamáres **4** · Sífnos · **3** Andíparos · Páros **5** · Kímolos **1** · Andímilos · **8** Adámas · Hóra **9** **1** Íos · **7** Mílos · Folégandros **5** · Síkinos **5** · Egiali · Amorgós · Léros · Kálymnos · Astypálea · Anáfi

See Mýkonos map
See Náxos map
See Santoríni map

Aegean Sea

0 km 40
0 miles

1	**Top 10 Sights** see pp88–91	
1	**Restaurants** see p93	
1	**Islands, Bays and Beaches** see p92	

8 Mýkonos

Ágios Sostis · Ágios Stefanos · **3** · **9** Fókos · Mýkonos Town · **4** **7** **8** Áno Méra · **1** **2** **3** **4** **5** **6** · **10** **2** **5** **7** **9** **10** · Platýs Gialós **4** · **8** Kalafáti · Ríneia · Delos **3**

0 km 5
0 miles

1	**Places to Eat and Drink: Mýkonos** see p95	
1	**Sights: Mýkonos** see p94	

capital, Hóra, is a collection of white houses, blue-domed churches and courtyards. Íos has some of the best beaches in the Cyclades. A short boat trip away is the remote island of Folégandros, popular with photographers due to its rugged beauty.

2 Náxos
MAP Q4

The centre of the Cycladic civilization during the Early Bronze Age (3000–2000 BC), Náxos has a phenomenal history. It was one of the first islands to use natural marble, a feature of the artifacts found at its archaeological sites and the material used for its Portára gateway, said to be the entrance to a 5th-century-BC Temple of Apollo. Náxos is the largest of the Cyclades, with its bustling capital the most populated. It is an island of citrus groves, beaches and sea caves.

Ruins of the Sanctuary of Apollo, Delos

3 Delos

An open-air museum, the uninhabited island of Delos is a UNESCO site in its entirety. It has fabulous archaeological remains dating mostly from Hellenistic times. The mythological birthplace of Apollo and Artemis, ancient Delos was a major place of worship, and remains one of Greece's foremost historic sites (see pp18–19).

Náxos

0 km   6
0 miles   6

Agiá
Ormos Abrám
Apóllon
Komiaki
Koronos
Náxos Town
Kinídaros
Melanés
Glinádo
Apeíranthos
Chalkí
Danakos
Áno Sagrí
Kanaki
Kastráki
Panermos
Agiassos
Kalantos

① **Places to Eat and Drink: Náxos**
see p97

① **Sights: Náxos**
see p96

Santoríni

0 km   5
0 miles   5

Baxédes
Koloumpos
Agía Eiríni
Oía
Thirassía
Korfos
Imerovígli
Kanakári
Néa Kaméni
Firá
Palaia Kaméni
Mesariá
Athiniós
Pýrgos
Megalochóri
Kamári
Emporió
Akrotíri
Périssa
Vlycháda

① **Places to Eat and Drink: Santoríni**
see p99

① **Sights: Santoríni**
see p98

Hillside village of Ano Sýros in Sýros

⑥ Ándros

MAP M4 ▪ Archaeological Museum: Plateía Kairi, Ándros Town, 22820 23664; open 9am–6pm Fri & Sun; adm

Life on this largely mountainous island revolves around the capital, Ándros Town, also known as Hóra, which boasts elegant Neo-Classical mansions in its centre, and a fascinating Archaeological Museum. The island's main settlements include Korthí, the inland medieval village of Mesariá, and Gávrio, while ancient Paleópolis lies near the west coast.

④ Sýros

MAP M6

A mountainous island interspersed with fertile plains and villages, cosmopolitan Sýros has fabulous beaches with amenities. Poseidonía and Vári are the liveliest tourist holiday spots, while Kíni and Galissás are quieter. Its capital is Ermoúpoli, which is built theatre-style around a deep harbour. In the 19th century Ermoúpoli was Greece's leading port with a thriving shipbuilding industry, which made Sýros one of the richest islands in the archipelagos.

⑤ Páros

MAP E4

Páros is a popular holiday destination and offers a huge choice of tavernas, restaurants and nightspots, especially in the towns of Paroikía, the capital, and Náoussa (see p52). Inland are tiny mountain villages, with Léfkes – a charming hamlet of medieval houses and alleyways – being the highest. Nearby Andíparos (see p92) centres on its Venetian Kástro and is popular for day trips from Páros.

⑦ Mílos

MAP D5

Mílos is famous for being where the *Venus de Milo* statue, now in the Louvre in Paris, was discovered. It is a volcanic island of horizontal white rock formations that have created an almost lunar landscape. Wealthy under the Minoans and Mycenaeans, and colonized by the Athenians in the 4th century BC, Mílos is an island of rural villages and beaches today. Its main town is Pláka, believed to be built on the acropolis of Mílos. About an hour away by boat lies Kímolos (see p92), home to the monk seal.

⑧ Mýkonos
MAP P6

Mýkonos is known for its beaches. Picturesque Psaroú and Eliá are two of the best and Paradise and Super Paradise (see p92) are livelier and popular with nudists. A party island, Mýkonos revolves around its capital town, an agreeable place of stone houses, many housing restaurants and bars. Visit Alefkándra, known as Little Venice, and the town's fine museums (see p94). Inland, the landscape is sprinkled with windmills and villages.

⑨ Santoríni
MAP T1

Crescent-shaped Santoríni owes its formation to a massive volcanic eruption that destroyed part of this once-circular island in 1450 BC. The resulting underwater crater has been associated with the legend of the lost city of Atlantis. Inhabited since Minoan times and named by the Venetians in the 13th century, this stunning island of dazzling white villages, museums, beaches, thermal springs and luxuriant vineyards is a quintessential Greek island.

⑩ Tínos
MAP N5

This holy island is famous for the church of the Panagía Evangelístria and its miraculous icon of the Virgin Mary, which attracts pilgrims throughout the year. Tínos is an unspoiled, fertile island. Its landscape is dotted with appealing villages, windmills, hundreds of chapels, and magnificent dovecotes.

Village of Mandrakia on Mílos

DAY TRIP FROM MÝKONOS TO DELOS

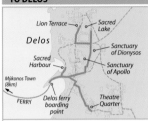

▶ MORNING

After a hearty breakfast at your accommodation, head down to the quayside at **Mýkonos Town** and look out for the departure points marked for **Delos** (see pp18–19). There are ferries and many smaller boats to choose from. Timings vary according to the season and the weather conditions, so be sure to check in advance to ensure you arrive in good time. Take refreshments with you as there is a limited choice on the boats and on Delos.

Most of the boats leave at around 10am for the 30-minute crossing. The sea can get a bit rough so be prepared. Once aboard, enjoy the sea breeze and your first glimpse of Delos ahead. On disembarking, head for the Maritime Quarter and see the **Sacred Harbour**, the **Theatre Quarter**, with its remains of mosaic-rich houses, the **Sacred Lake**, with its iconic terrace of lion statues, and the **Sanctuaries of Apollo and Dionysos**. There is a small restaurant in the Sanctuary of Apollo for a quick bite before you head back to Mýkonos.

AFTERNOON

Boats tend to leave Delos at 12:30pm, 1:30pm and 3pm, but again be sure to check as timings can vary. If you aim for the 3pm crossing it will give you plenty of time to see the extensive archaeological site on Delos and absorb its atmosphere. Once back on Mýkonos, enjoy a leisurely afternoon siesta before a delicious meal at a taverna (see p95) on the harbourside.

See map on pp88–9 ←

Islands, Bays and Beaches

1 Kímolos Island, Mílos
MAP E4

Taking its name from *kimólia*, meaning chalk, which is mined here, this volcanic island is home to a few small communities and good tavernas. Beaches and caves make it a popular day-trip destination.

2 Ágios Geórgios Beach, Santoríni
MAP V3 ■ Southeast coast

A black-sand beach, Ágios Geórgios has lots of tavernas and bars, plus sunbeds to rent. There's also scuba diving and windsurfing *(see p48)*.

3 Andíparos Island, Páros
MAP E4

This unspoiled island has beautiful beaches and bays. Its tavernas are welcoming and its cafés are chic. A range of activities are offered.

4 Agrári Beach, Mýkonos
MAP P6 ■ South coast

This attractive sandy beach, enclosed by rocks, is an ideal setting in which to relax. Its azure water is safe and there is a taverna and bar *(see p50)*.

5 Síkinos Island, Íos
MAP E5

This rugged island, with deep bays and pretty beaches, is known for squid-fishing, sweet wine and aromatic thyme honey. Hóra, its main town, has lovely cobbled streets.

6 Golden Beach, Ándros
MAP M4 ■ Batsí

One of the best beaches on Ándros, this long stretch of sand lies next to Batsí resort. It has shallow water and all amenities for visitors.

7 Pahiá Ámmos Beach, Tínos
MAP P5 ■ Near Ágios Ioánnis

Set in a natural bay, this beach is made up of endless sand dunes. Surrounded by a rugged landscape, it is isolated and peaceful *(see p50)*.

8 Fínikas Bay, Sýros
MAP M6 ■ Southwest coast

Picturesque and bustling with fishing boats and yachts, this sheltered bay has a long sandy beach. Its barren backdrop is dotted with pine trees and white houses, many of which house fish tavernas *(see p49)*.

9 Super Paradise Beach, Mýkonos
MAP P6 ■ South coast

This long beach, boasting crystal-clear water and many tavernas, is famous for its party atmosphere. It is popular with nudist and gay holiday-makers.

10 Agathopés Beach, Sýros
MAP M6 ■ Near Poseidonía

A natural habitat for flowers and birds, protected by the Ministry of Environment, this sandy beach has an offshore islet and safe swimming.

Alopronia harbour, Síkinos

Restaurants

1 Symposion, Tínos
MAP N5 ▪ Evagelistrias 13, Tínos Town ▪ 22830 24368 ▪ €€€

In an elegant building close to the harbour, Symposion is a landmark restaurant. It is all about crisp linens, sparkling wine glasses, gourmet-style Greek dishes and attentive staff.

2 Levantis Restaurant, Páros
MAP E4 ▪ Párikia ▪ 22840 23613 ▪ €€

Delicious Mediterranean and Asian dishes are on offer at this little place in the Old Town. Try the rabbit in yogurt with olives and aubergine.

Levantis Restaurant, Párikia Old Town

3 Allou Yialou, Sýros
MAP N6 ▪ Kíni ▪ 22810 71196 ▪ €€

Decorated in a nautical style, this waterside taverna is known for its artful fish platters, delicious desserts and fabulous views of the sun as its sets over the sea.

4 Delfini Restaurant, Sífnos
MAP E4 ▪ Kamáres ▪ 22840 33740 ▪ €

Overlooking the port and bay, this quintessential blue and white taverna has an innovative Mediterranean meat, vegetable and fish menu.

5 Kritikos, Folégandros
MAP E5 ▪ Piátsa Square, Hóra ▪ 22860 41219 ▪ €€

This traditional taverna, more than 30 years old, is well-known for its spinach, pumpkin and cheese pies.

PRICE CATEGORIES
For a three-course meal for one with half a bottle of wine (or equivalent meal), taxes and extra charges.
....
€ under €30   €€ €30–€50   €€€ over €50

---

6 Takis Restaurant, Sérifos
MAP E4 ▪ Livádi ▪ 22810 51159 ▪ €€

Located on the tree-lined waterfront, Takis serves Greek dishes such as *kléftiko* and *souvláki (see p62)* from an extensive menu. Wines focus on vintages from the islands.

7 Byzantio Restaurant, Kýthnos
MAP E4 ▪ Mérichas ▪ 22810 32259 ▪ €

Overlooking the bay with tables on the beach in summer, this attractive taverna serves traditional Cycladic cuisine. Meats are cooked on charcoal to age-old recipes.

8 Aragosta Restaurant, Mílos
MAP D5 ▪ Adámas ▪ 22870 22292 ▪ €€

This elegant restaurant, which is housed in a Neo-Classical mansion, offers an extensive Greek and international menu. Grills are a speciality, along with *tyropitákia* (cheese pies) and local seafood.

9 Lord Byron, Íos
MAP E5 ▪ Hóra ▪ 22860 92125 ▪ €€

Mediterranean cuisine, fine wines and jazz music give the Lord Byron a sophisticated edge. The menu features Íos lamb with honey, seafood and pasta, plus homemade desserts.

10 Dio Horia, Tínos
MAP N5 ▪ Dio Horia Village ▪ 22830 41615 ▪ €€

Located in the hills of Tínos Island, Dio Horia has a dining terrace where great views of the harbour can be savoured over an *ouzo*. The cuisine is classic Greek with dishes like *souvláki* on the menu.

See map on pp88–9 ←

Sights: Mýkonos

① Alefkándra
MAP P6 ▪ Mýkonos Town

With colourful 16th- and 17th-century pirates' houses built on the water's edge, Alefkándra, also called Little Venice, is the artists' quarter of Mýkonos Town.

② Archaeological Museum
MAP P6 ▪ Harbourside, Mýkonos Town ▪ 22890 22325 ▪ Open 3–10pm Mon, 9am–10pm Tue & Thu–Sun, 9am–4pm Wed ▪ Adm

Archaeological Museum, Mýkonos

Housed in a beautiful Neo-Classical building, the Archaeological Museum has a fine collection of jewellery and funerary statues.

③ The Harbour
MAP P6 ▪ Mýkonos Town

Home to the island's official pelican mascot, Petros Peter, the harbour is dotted with tavernas and white reed-roofed 16th-century windmills.

Mýkonos Town's harbour

④ Panagía Paraportianí
MAP P6 ▪ Kástro, Mýkonos Town

This 15th-century church, with four chapels and a fifth one built on top, is located on a medieval fortress's gate.

⑤ Léna's House
MAP P6 ▪ Tría Pigádia, Mýkonos Town ▪ 69774 39832 ▪ Open Apr–Oct: 6:30–9:30pm daily

This 19th-century mansion was the home of local resident Léna Skivánou. Now a museum, the building has been beautifully restored.

⑥ Folk Museum
MAP P6 ▪ Kástro, Mýkonos Town ▪ 69774 39832 ▪ Open Apr–Oct: 4:30–8:30pm Mon–Sat

This museum has displays of rare traditional textiles, ceramics, historical photographs and furnishings.

⑦ Áno Méra
MAP P6 ▪ Central Mýkonos

Stone houses covered with bougainvillea line Áno Méra's alleyways and the monastery of Panagía Tourlianí dominates the skyline.

⑧ Moní Panagía Tourlianí
MAP P6 ▪ Central Mýkonos ▪ 22890 71249 ▪ Open by arrangement

Founded in the 16th century and restored in 1767, this monastery is dedicated to the island's protectress. Its white exterior, red-domed roof and a sculpted marble tower are the work of Tíniot craftsmen.

⑨ Paleokástro
MAP P6 ▪ Central Mýkonos

Believed to be the site of an ancient city, Paleokástro lies on a hill in the island's most verdant area.

⑩ Aegean Maritime Museum
MAP P6 ▪ Tría Pigádia, Mýkonos Town ▪ 21081 25547 ▪ Open Apr–Oct: 10:30am–1pm and 6:30pm–9pm daily; Nov–Mar: 8:30am–3pm daily ▪ Adm

This museum explores Greek maritime history, with a particular focus on merchant ships (see p42).

Places to Eat and Drink: Mýkonos

1 Tasos Restaurant
MAP P6 ■ Parága beach ■ 22890 23002 ■ €€

A favourite with locals, this restaurant offers fine fish and seafood dishes – octopus cooked on charcoal is a speciality. The terrace overlooks the bay, and local wines are served.

2 Skandinavian Bar, Mýkonos
MAP P6 ■ Mýkonos Town ■ 22890 22669 ■ €€€

This is one of the best nightspots in Mýkonos to enjoy music, dancing and drinks at the best prices (see p60).

3 Kiku
MAP P6 ■ Harbourside, Mýkonos Town ■ 22890 20100 ■ €€

This restaurant's Japanese menu provides a delicious alternative to local cuisine. Fresh fish is cooked to authentic oriental recipes.

4 La Cucina di Daniele
MAP P6 ■ Áno Méra ■ 22890 71513 ■ €€

With its colourful decor and lively atmosphere, this restaurant is a great place to enjoy homemade pasta dishes. The Italian owner recommends wine based on the day's menu.

5 Avra Restaurant
MAP P6 ■ Kalogéra Street, Mýkonos Town ■ 22890 22298 ■ €€

This attractive restaurant offers intimate indoor and outside dining. A tasty salad and appetizer menu is complemented by mains plus an à la carte selection.

6 Fokos Taverna
MAP P6 ■ Fókos beach ■ 69446 44343 ■ €

With a menu featuring aubergine salad, mussels in wine and delicious desserts, Fokos is located in a traditional house. Drinks include fine wines, speciality coffees and oúzo.

PRICE CATEGORIES

For a three-course meal for one with half a bottle of wine (or equivalent meal), taxes and extra charges.

€ under €30   €€ €30–€50   €€€ over €50

7 Sea Satin Market
MAP P6 ■ Alefkándra, Mýkonos Town ■ 22890 24676 ■ €€

This trendy eatery comes alive after sunset as the lights go on. Fresh fish is cooked on charcoal grills (see p65).

Fresh fish, a speciality at Spilia

8 Spilia Seaside Restaurant & Cocktail Bar
MAP P6 ■ Seafront, Kalafáti ■ 69494 49729 ■ €€

Diners at this elegant beachside taverna choose the fish for their meal from a water pool replenished by the sea. The menu has Italian flair.

9 La Maison de Catherine
MAP P6 ■ Nikíou 1, Ag. Gerasímou, Mýkonos Town ■ 22890 22169 ■ €€€

This fashionable restaurant offers classic Greek dishes with a French twist. The wines are great (see p64).

10 Raya Restaurant
MAP P6 ■ Gialós waterfront ■ 22890 77766 ■ €€

This stylish lounge bar and restaurant serves breakfasts of fresh juices and feta omelettes as well as lunches, dinners, cocktails and coffees.

See map on pp88–9

Sights: Náxos

Fresco inside Chalkí church

1 Chalkí
MAP R5

The former capital of Náxos, Chalkí is full of Neo-Classical buildings. It is surrounded by olive groves and its 9th-century church has notable frescoes and well-preserved towers.

2 Náxos Town
MAP Q4

Dominated by the Portára gateway, Náxos Town is a heady mix of history and modernity. Its Venetian castle and the church of Panagía Myrtidiótissa are well worth a visit.

3 Apeíranthos
MAP R5

White marble houses dot the hillside of this pretty village. Sights of interest include one of the island's oldest churches, the Panagía Apeíranthos, and the small archaelogical and geological museums

4 Mount Zeus
MAP R6

Also known as Mount Zía or Zía Óros, this is the Cyclades' highest mountain at more than 1,000 m (3,300 ft). It is well known for its hiking trails.

Trekking the rocky path to the summit of Mount Zeus

5 Glinádo
MAP Q5

A charming village of whitewashed stone houses with blue painted doors and tiny courtyards full of bougain-villea, Glinádo has been preserved as a traditional Náxiot community.

6 Apóllon
MAP S4

The fishing village of Apóllon, or Apóllonas, is the site of an ancient marble quarry. Its entrance has a huge 6th-century-BC statue, Kouros. Tavernas line its sandy beach.

7 Komiakí
MAP R4

Komiakí, or Koronída, is one of the island's highest villages and is known for traditional crafts, especially lace.

8 Melanés Valley
MAP R5

Full of olive and citrus trees, this valley has pretty villages, including Kourounchóri, Flerió and Melanés.

9 Tragéa Valley
MAP R5

The Panagía Drosianí church is one of this valley's most memorable sights. The many groves attest to Tragéa's flourishing olive industry.

10 Áno Sagrí
MAP R5

Windmills and Byzantine churches dot the lush plateau around Áno Sagrí. Its chapel of Ágios Ioánnis Gyroulás is said to be built over the 530 BC Temple of Demeter.

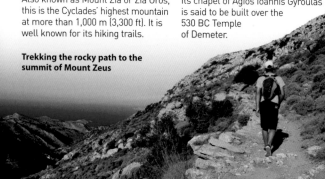

Places to Eat and Drink: Náxos

PRICE CATEGORIES

For a three-course meal for one with half a bottle of wine (or equivalent meal), taxes and extra charges.

€ under €30   €€ €30–€50   €€€ over €50

1 Scirocco Restaurant
MAP Q4 ▪ Náxos Town ▪ 22850 25931 ▪ €€

Lemonáto, a dish of lamb in lemon, and Katerina's chicken, cooked with mustard and orange, feature on the menu of this popular eatery. It is open from breakfast to dinner.

2 Kavouri Restaurant
MAP Q5 ▪ Ágios Geórgios beach ▪ 22850 27276 ▪ €

A landmark since 1955, this popular restaurant has traditional Greek decor and is near the beach. It serves classic dishes and tasty fresh fish.

3 Apolafsi Restaurant
MAP Q4 ▪ Kastraki ▪ 22850 75483 ▪ €

Shrimp balls and veal stroganoff are among the 100 or so dishes on this attractive waterside restaurant's menu. A large wine list and live entertainment add to the experience.

4 I Avli
MAP Q5 ▪ Agía Anna ▪ 69774 44123 ▪ €€

With its fresh blue and white decor and fish drying from the sea drying on the terrace, this restaurant is one of a few in the sleepy resort of Agía Anna. Fish grilled or cooked slowly in a sauce are the specialities.

5 Nostimon Hellas
MAP Q4 ▪ Ioannou Paparigopoulou & Tripodon, Ágios Geórgios ▪ 22850 25811 ▪ €€

A charming paved courtyard awash with colour from bougainvillea plants provides the setting for a menu of Greek favourites. Most dishes are *scháras*, meaning from the grill. The signature dish is veal with rosemary.

6 To Souvlaki tou Maki
MAP Q4 ▪ Papavasiliou, Náxos Town ▪ 22850 26002 ▪ €€

Memorabilia-covered stone walls and a reed-style ceiling give To Souvlaki tou Maki a traditional taverna feel. Dishes are hearty Náxian and Greek.

7 Yialos Restaurant
MAP Q4 ▪ Ágios Geórgios beach, Náxos Town ▪ 22850 25102 ▪ €€

Close to the beach, this taverna serves cool drinks, snacks, and meat and fish straight off the charcoal grill.

Squid, popular on Greek menus

8 Picasso Mexican Bistro
MAP Q4 ▪ Plaka Beach ▪ 22850 41188 ▪ €€

Quesadillas filled with meat and beans, plus tacos, fajitas and burritos, are served with frozen Margaritas on the beach.

9 Kontos Restaurant
MAP Q5 ▪ Mikrí Vígla ▪ 22850 75278 ▪ €

Apart from growing its own fruit and vegetables, this lovely open-air restaurant also rears its own animals to produce the freshest of Náxiot dishes.

10 Oasis
MAP Q4 ▪ Ágios Geórgios ▪ 22850 25606 ▪ €

Eat under the shade of citrus trees in the courtyard and choose from Greek dishes that use produce from local sources or fish fresh from the sea.

See map on pp88–9

Sights: Santoríni

1 Firá
MAP U2

The island's capital, Firá, also called Thíra, has beautiful Venetian buildings that survived the 1956 earthquake.

2 Ancient Thíra
MAP V3 ▪ Mésa Vounó peninsula

The site of the ancient Dorian city of Thíra can be found laid out on terraces at the end of the Mésa Vounó peninsula.

3 Pýrgos
MAP U2

A hilltop village of blue and white houses built around a Venetian castle, Pýrgos is home to the land-mark Monastery of Prophítis Ilías.

4 Oía
MAP U1

Much photographed due to its blue and white houses, Oía is also famous for its sunsets *(see p53)*. Near the port is a small maritime museum.

Pot, Museum of Prehistoric Thíra

5 Museum of Prehistoric Thíra
MAP U2 ▪ Firá ▪ 22860 23217 ▪ Open 8:30am–3pm Tue–Sun ▪ Adm

Fascinating exhibits displayed here include fossilized olive leaves and clay pots, plates, tools and other artifacts discovered at ancient Thíra.

6 Kaméni Islets
MAP U2

These two uninhabited islets, Néa Kaméni and Palaia Kaméni, first appeared in the sea after a volcanic eruption in 1450 BC. They were formed from lava and ash – and remain volcanic to the present day.

7 Imerovígli
MAP U2

This quiet, traditional village provides visitors with a rare glimpse into everyday life. It has stone houses and chapels and great views of the flooded caldera *(see p55)*.

8 Archaeological Museum
MAP U2 ▪ Firá ▪ 22860 22217 ▪ Open 8am–3pm Tue–Sun ▪ Adm

Sculptures and inscriptions from Archaic to Roman times and vases and clay figurines from Geometric to Hellenistic eras tell the story of Santoríni's fascinating past.

9 Ancient Akrotíri
MAP U3 ▪ Akrotíri ▪ 22860 81939 ▪ Open 8am–5pm daily

This late Neolithic-era city is among Greece's most important. Look out for the frescos of two boys boxing and the voluptuous dark ladies.

10 Thirassía Island
MAP T2

Separated from Santoríni by volcanic activity, Thirassía is a rocky island dotted with small hamlets, such as Manolás, and a few tavernas.

Oía, trailing down the mountain

Places to Eat and Drink: Santoríni

1 Mario Restaurant

MAP V2 ■ Agía Paraskeví, Monólithos ■ 22860 32000 ■ €€

Baked aubergine and seafood pasta dishes are among the delicacies at this chic, long-established taverna located right beside the beach.

2 Ambrosia Restaurant

MAP U1 ■ Oía ■ 22860 71413 ■ €€

Ambrosia is known for its elegant decor, a candle-lit terrace looking over the volcano and great Mediterranean dishes. Booking is advisable.

3 Blue Note Restaurant

MAP U2 ■ Imerovígli ■ 22860 23771 ■ €€

Tucked away in an old stone building, this restaurant, with a terrace overlooking the bay, specializes in fish.

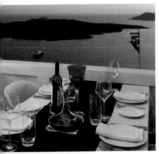

View from the terrace, Koukoumavlos

4 Koukoumavlos

MAP U2 ■ Firá ■ 22860 23807 ■ €€€

Dine on gastronomic-quality food among enchanting period decor. The food is exquisite, and pricey.

5 Lithos Restaurant

MAP U2 ■ Firá ■ 22860 24421 ■ €€

Housed in a period building on the harbourside, the Mediterranean menu here features fish platters and grills.

PRICE CATEGORIES

For a three-course meal for one with half a bottle of wine (or equivalent meal), taxes and extra charges.

€ under €30 **€€** €30–€50 **€€€** over €50

6 Mama Thira Taverna

MAP U2 ■ Harbourside, Firostefáni, Firá ■ 22860 22189 ■ €€

The dining hall of Mama Thira has nostalgic items adorning its walls and a terrace overlooking the bay. The Mediterranean cuisine and wine is served with exceptional flair.

7 To Pinakio

MAP V3 ■ Kamári beach ■ 22860 32280 ■ €€

A locals' favourite, To Pinakio serves wholesome Greek cuisine. Octopus and squid, cooked with local herbs, and its speciality *moussakás* are menu classics. Desserts are homemade.

8 Lauda Restaurant

MAP U1 ■ Oía ■ 22860 72182 ■ €€€

Food at Lauda is, quite simply, a work of art. Inside the Andronis Boutique Hotel, the dining area overlooks the tiny lanes of Oía. Sip cocktails while watching the sunset.

9 Mylos Bar & Restaurant, Santoríni

MAP E5 ■ Firostefani ■ 22860 25640 ■ €€€

With spectacular views of the caldera and the Aegean, this stylish restaurant serves modern gastronomic delights. Be sure to leave room for one of the mouthwatering desserts *(see p60)*.

10 Melitini, Santorini

MAP T1 ■ Oía ■ 22860 72343 ■ €€

This friendly eatery serves traditional tapas on a rooftop terrace that overlooks Santorini caldera in the Aegean Sea.

See map on pp88–9

🔟 Crete

Crete (*see pp26–31*) has always been an island that captures the imagination, with some of the world's finest ancient sites, including the palaces of Phaestos and Knossos. The birthplace of Europe's oldest civilization, the enigmatic Minoan culture that flourished over 4,000 years ago, Crete was also ruled by the Romans, Byzantines and Venetians. Cretan resistance against occupation by the Ottoman Turks (1669–1898) and the Germans in World War II is legendary. The largest of the Greek Islands, mountainous Crete has a distinct culture and its own dialect. Its capital city is the cosmopolitan Irákleio, also known as Heraklion. Chaniá and Réthymno are its other two major towns, while the Samariá Gorge is its top natural attraction.

Ruins of the main structure, Palace of Knossos

1 Palace of Knossos

Knossos was the capital of Minoan Crete and its impressive palace was an architectural marvel. The original palace, dating from about 1900 BC, was destroyed by an earthquake but was quickly rebuilt. The remains were discovered and restored by Sir Arthur Evans in 1878 (*see pp30–31*).

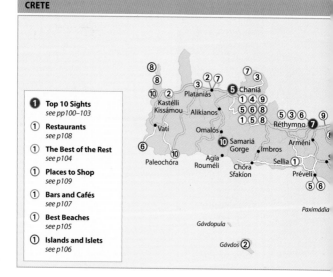

CRETE

2 Phaestos Palace

One of the great city-state palaces of the Minoan period, this fabulous structure lay hidden for over 3,000 years before being discovered by Italian archaeologist Frederico Halbherr in the late 19th century. The well-preserved sights include the grand staircase and the central courtyard (see p27).

3 Amári Valley and the Agía Triáda

MAP E6

Scattered with traditional villages, chapels and cherry orchards, the Amári Valley is an unspoiled area of mountains and valleys that is dominated by Mount Idi at 2456 m (8080 ft). From here, you can visit Phaestos and the nearby Royal Villa at Agía Triáda. Built in the 16th century BC, this Minoan villa was one of the first palaces to be built with much smaller proportions than those of Phaestos. The structure consisted of two wings with court-yards and workshops. Excavations unearthed tablets with Minoan script, a Harvester vase and rhyton jug, sug-gesting great wealth (see pp26–7).

Ruins of the ancient temple, Górtys

4 Górtys

Thought to have been inhabited since Neolithic times, Górtys, or Gortyn, became the powerful capital of Crete, ahead of Phaestos, after the Roman invasion of 65 BC. Its power lasted until AD 7, when invading Arabs destroyed the city. Located in the village of Ágii Déka, the remains of its citadel and agora (market place), the *praetorium* (governor's house), temples and a cemetery can be visited. A three-aisled basilica, Ágios Títos, domi-nates the entrance. The *odeion* (small theatre) ruins are famous for their inscriptions (see p27).

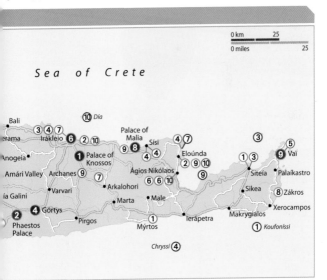

Mountains provide a backdrop for the quayside at Chaniá

5 Chaniá
MAP D6

A sprawling fortified city set in the foothills of the Lefká Óri mountains, Chaniá was one of the most powerful cities during the Minoan period. Decline was followed by renewed importance under the Venetians. Chaniá was the capital of Crete until 1971. Today, life revolves around its scenic harbour and the Spiántza Quarter, with its Venetian houses lining cobbled streets, modern restaurants and art galleries.

6 Irákleio
MAP E6

Crete's capital city is a mixture of Venetian mansions, busy streets, a harbour full of fishing boats and squares full of chic restaurants, tavernas and designer fashion shops. A traditional Greek city with a distinctly modern twist, at its heart is the pedestrianized Plateía Eleftheríou Venizélou, with the 13th-century

Vase, Irákleio Archaeological Museum

church of Ágios Márkos. Must-see sights include the Archaeological Museum, which boasts an impressive collection of Minoan art (see pp27–8).

7 Réthymno
MAP E6 ■ Historical and Folk Art Museum: Vernadou; 28310 23398; open 10am–2:30pm Mon–Sat; adm

With a well-preserved Venetian fortress, picturesque harbour famous for its 13th-century lighthouse, and fish tavernas, Réthymno is a refined holiday destination. One of its Venetian mansions now houses the small Historical and Folk Art Museum. Once the Greco-Roman city of Rithymna, Réthymno at one time was important enough to mint its own coins. The town's crest is based on one of these old coins.

8 Palace of Malia
A little inland are the archaeological remains of Malia, one of the island's most prosperous Minoan cities. The palace, noted for its layout, was constructed around a courtyard with a central altar. It has raised paths, known as processional ways, leading to its west entrance. Inside there is a labyrinth of crypts that were probably used for cult worship. Treasures unearthed here include the famous *kernos*, believed to be a vessel used for sacrifices (see pp26–7).

GÓRTYS LAW CODE

Inscribed on the north wall of Górtys Odeion, the Law Code related to all domestic matters, from marriage to the division of property. The script, known as boustrophedon, has alternate lines written from right to left, then left to right. It uses symbols of the Doric Cretan dialect, dating from around 500 BC.

⑨ Vaï

MAP F6 ■ Near Palaiokastró

There are few places in Europe that could be called "unique", but Vaï is arguably one of them. Covering a vast area, it is a dense forest made up of *phoenix theophrasti* (Cretan Date), palm trees that have grown over time naturally to become the largest in Europe. The forest, which is in a protective enclosure and designated an Aesthetic Forest, lies right beside a long stretch of beach on the eastern tip of Crete where there is a natural stream fed by a spring. The beach itself is now one of the largest tourist attractions on the island of Crete and has, sadly, become very commercialized.

⑩ Samariá Gorge

Created over millennia by a river running between two mountains, the dramatic Samariá Gorge is one of the longest and deepest canyons in Europe. Crete's most famous natural wonder lies 44 km (27 miles) south of Chaniá. Extremely popular with hikers, visitors can take the fairly strenous walk from the village of Omalós, along the 16-km- (10-mile-) long gorge to the coastal village of Agía Rouméli. Along the way are towering rock faces, springs and chapels, such as Ágios Nikólaos. The area was designated a national park in 1962, primarily to protect the rare kri-kri goat *(see p53)* and its natural habitat *(see p27)*.

Steep footpath at Samariá Gorge

See map on pp100–101 →

A HIKE ALONG THE SAMARIÁ GORGE

▶ Have a hearty breakfast before you leave and pack plenty of snacks, a water bottle (which can be replenished in the gorge), sun cream and plasters before heading off on this approximately 6-hour trip exploring the **Samariá Gorge**. It is best to wear light clothing but take a jumper, as it can be cold at higher altitudes, and always wear a hat. You will need sturdy walking shoes as the path is mainly made up of rocks and pebbles.

Once fully prepared, head for the village of **Omalós**. The best option, if you plan to hike the length of the gorge, is to go by bus because you'll be tired going home, but if you intend hiking just a short distance and returning to Omalós, there is a car park. Opening times can vary according to the weather, so be sure to check. Aim to arrive early.

From Omalós, make your way across the plateau before taking a sharp descent down the snaking path known as **Xylóskalo**. It is around 2 km (1 mile) long. Follow the signposted route, which will take you past charming chapels like the tiny **Ágios Nikólaos** and **Óssia Maria**, which has some fine 14th-century frescoes. You will also pass the deserted village of **Samariá** and a landmark spot known as **Sideróportes**, where the path considerably narrows between towering mountainsides. Then heading down the Gorge, you will pass by the village of **Old Agía Rouméli**, before reaching the new **Agía Rouméli** located on the coast, about 6 hours after you started. Now take a ferry to **Chóra** Sfakíon, where there are buses.

The Best of the Rest

 Siteía
MAP F6 ■ Northeast coast

This coastal village is surrounded by vineyards and is known for its quality Cretan wines. See its Venetian fort, now a cultural venue, its historic harbour and its Victorian mansions.

 Kastélli Kissámou
MAP D6 ■ West coast

This quiet town with a handful of tavernas is a great base for exploring the unspoiled Gramvoússa peninsula. Nearby is the superb beach and ancient site of Falásarna.

 Akrotíri Peninsula
MAP D6 ■ Near Chaniá

The Akrotíri Peninsula is a barren place and a hiker and wildlife enthusiast's paradise. A shrine to Crete's national hero, Elfthérios Venizélos, sits on a hill, while tiny monasteries dot its landscape.

Ruins on Spinalónga island

 Spinalónga
MAP F6 ■ Northeast coast

This heavily fortified islet and former leper colony makes a dramatic sight from the ferries crossing from Eloúnda. It's been home to Olouites, Venetians and Ottomans *(see p106)*.

 Préveli
MAP E6 ■ South coast

The site of two monasteries, the 18th-century Moní Préveli and the earlier Moní Agíou Ioánnou, Préveli is a small village reached via the Kourtaliótiko gorge.

The attractive town of Ágios Nikólaos

 Ágios Nikólaos
MAP F6 ■ Northeast coast

Overlooking the Mirabéllou bay, Ágios Nikólaos has grown from a small Venetian fishing village to a cosmopolitan holiday resort. In Hellenistic times it was a city known as Lato.

 Mátala
MAP E6 ■ South coast

Mátala is a bustling holiday resort with a fabulous history. It is said to be where St Paul landed on Crete and where Menelaos, the husband of Helen of Troy, was shipwrecked.

 Zákros
MAP F6 ■ East coast

On the slopes of Mount Deléna, Zákros is an unassuming coastal village. It gained prominence in 1961 when the remains of a Minoan palace with artifacts were discovered.

 Eloúnda
MAP F6 ■ Northeast coast

With its popular sandy beaches and coves, Eloúnda is a busy holiday destination. Site of the ancient city-state Olous, it was fortified under Venetian rule. An isthmus links it to the Spinalónga peninsula.

 Paleochóra
MAP D6 ■ West coast

Paleochóra's landmark is the remains of a Venetian castle destroyed by pirates. The village is on a headland that divides the beach in two. The resort is popular with windsurfers.

Best Beaches

1 Mýrtos Beach
MAP E6 ■ South coast

Surrounded by citrus groves and banana plantations, the secluded beach of Mýrtos is one of the most idyllic spots on the island. It is sandy, with a few pebbles here and there.

2 Amníssos Beach
MAP E6

The main beach of Irákleio is at Amníssos, a stretch of golden sand that has lots of charm. It is famous for having been the port of Knossos, and many archaeological remains have been discovered here.

3 Siteía Beach
MAP F6 ■ Northeast coast

Awarded a Blue Flag for cleanliness, this sandy beach lines the edge of a deep, horseshoe-shaped bay and is near the picturesque Siteía village. It is popular for windsurfing.

4 Sísi's Beaches
MAP E6 ■ East coast

Sísi village has two beaches, one a secluded stretch of sand in the small Avláki bay and the other just off the harbour near tavernas and bars.

5 Vaï Beach
MAP F6 ■ East coast

With its wild date-palm grove and its golden sand dunes, Vaï beach is among the island's most beautiful *(see p49)*.

Visitors relaxing on Vaï beach

6 Préveli Beach
MAP E6 ■ South coast

The Kourtaliótis river meets the sea here at Préveli, where the green river, blue sea, date palms and "Greek bamboo" (calamus reeds) create a tropical oasis feel.

7 Stavros Beach
MAP D6 ■ North coast

Tucked within a sandy cove, this unspoiled beach is best known as one of the locations for the film *Zorba the Greek*. There are only a few small tavernas for refreshment.

8 Gramvoússa Peninsula
MAP D6 ■ West coast

Along with Falásarna, there are many isolated and peaceful beaches and sandy coves dotting the coast of this unspoiled peninsula. Most are only accessible by boat.

9 Réthymno Beach
MAP E6 ■ North coast

Although usually crowded because its main section is in Réthymno city, this beach is nonetheless beautiful. It is about 20 km (13 miles) long and dotted with palm trees and tavernas.

10 Falásarna Beach
MAP D6 ■ West coast

In the far northwest of Crete, this is one of the cleanest beaches on the island. The long stretch of sand is lapped by turquoise sea. It is known for its sunsets. Nearby tavernas provide refreshment *(see p49)*.

Islands and Islets

1 Koufoníssi Island
MAP F6 ■ Off Goúdouras

Characterized by its desert landscape of fine sand dotted with tamarisk trees, this tiny deserted island has archaeological remains that suggest it was inhabited in ancient times.

View out to sea, Koufoníssi island

2 Gávdos Island
MAP D6 ■ Off Sfakiá

This cedar- and pine-covered island is Greece's southernmost boundary. It has quaint villages and beaches. According to Homer's *The Odyssey*, the nymph Calypso lived here.

3 Dragonáda Island
MAP F6 ■ Off Siteía

Remains of ancient habitation and early Christian tombstones have been found here in the Dionysádes archipelago. Now uninhabited, it is covered with vegetation.

4 Chryssí Island
MAP E6 ■ Off Ierápetra

An island of volcanic rock and sand dunes, Chryssí is uninhabited yet it has a well-preserved 700-year-old church, an ancient port and remains from the Minoan period.

5 Paximádia Island
MAP E6 ■ Off Mátala

These small rocky islets were once named the Islets of Dionýsoi, after Dionysos, the god of wine. Today, they are a popular boat-trip from Mátala.

6 Elafoníssi Islet
MAP D6 ■ Off Paleochóra

This scenic islet is covered with trees and has beaches made up of tiny coral fragments, which give them a pinkish hue. It is surrounded by shallow aquamarine water.

7 Spinalónga Island

Dominated by an imposing Venetian fortification, this lush island makes a popular day-trip spot, in part due to the popularity of Victoria Hislop's novel *The Island*, which centres on its history as a leper colony. Ferry boats dock at its picturesque quayside, letting visitors explore the island *(see p104)*.

8 Gramvoússa Island
MAP D6 ■ Off Kastélli Kissámou

This small island off the Gramvoússa peninsula is a haven for wildlife. In Venetian times it was a key defence location. The former stronghold now attracts day-trippers in summer.

9 Psíra Island
MAP F6 ■ Off Ágios Nikólaos

This barren island traces its history back to Minoan times, when it was a hub for fishing and agriculture. It is believed that there was a harbour here lined with merchants' homes.

10 Día Island
MAP E6 ■ Off Irákleio

This island is a nature reserve that is a habitat for the rare Cretan goat, the kri-kri *(see p53)*. It boasts great waters for swimming and snorkelling, coves to explore and a church to visit.

Kri-kri goat, found on Día Island

Bars and Cafés

An array of Scandinavian craft beer on offer at the bar at Klik

1 Klik Bar
MAP D6 ▪ Old Venetian Port, Chaniá ▪ 69459 32971
Every night is a party at the Klik for the young and trendy who dance to R&B, indie, hip-hop and rock.

2 Cosmos Piano Bar
MAP D6 ▪ Páno Plataniás, Chaniá ▪ 28210 68558
A stylish venue overlooking the bay, this piano bar serves a huge selection of wines, cognacs and malts, along with gourmet delicacies.

3 Maleme Tavern Snack Bar
MAP D6 ▪ Maleme, Chaniá ▪ 28210 62049
Pancakes and coffee are served for breakfast and light Cretan dishes through the day at this bar near the beach. Enjoy a cocktail in the evening.

4 Jolly Roger, Crete
MAP E6 ▪ Sísi ▪ 28410 71656
Located in the pretty harbour village of Sísi, Jolly Roger is the perfect place to relax with a cocktail or take part in the variety of entertainment on offer here, such as live music and quiz nights *(see p60)*.

5 Peacock Tail Bar
MAP D6 ▪ Kanevaro 7a, Chaniá ▪ 69703 40350
With modern stone decor inside and a pleasant outdoor seating area, Peacock Trail is a great place to sip lavish cocktails and listen to music.

6 Chaplin's
MAP E6 ▪ Elfthérios Venizélos 52, Réthymno ▪ 28310 24566
This rock music-themed pub near the old harbour serves food and beers all day long. Sit inside or prop up the streetside bar.

7 La Brasserie
MAP E6 ▪ Karai 15, Irakleio ▪ 28140 01418
The decor and bistro-style menu take influences from around the world at this venue. Choose from a wide range of cocktails and time your visit for a live jazz or Cuban-themed night.

8 Sinagogi Bar
MAP D6 ▪ Kondilaki, Chaniá ▪ 28210 95242
A beautiful old Cretan building with stonework, arches and terraces, on a small square near the seafront, is the setting for Sinagogi.

9 Slainte Irish Bar
MAP E6 ▪ Sokrátous 2, Stális ▪ 28970 32879
Guinness features here and there are also cocktails and a menu of hearty meals to enjoy while watching sports on big screens. Bands play regularly.

10 The Whisky Bar
MAP E6 ▪ Seafront, Kokkiní Hani ▪ 28107 61563
Enjoy Guinness, whiskies, beers and cocktails at this seafront bar. There are panoramic views by day and sunsets in the evening.

See map on pp100–101

Restaurants

1 Theodosi
MAP D6 ▪ Paparigopoulou 99, Chaniá ▪ 28210 93733 ▪ €€€

Upmarket Theodosi has elegant decor with crisp linens, and the à la carte menu celebrates Cretan cuisine with dishes given a creative twist *(see p65)*.

2 Marilena Restaurant
MAP F6 ▪ Harbourside, Eloúnda ▪ 28410 41322 ▪ €€

Enjoy flambéed dishes, fresh fish and seafood or try Greek *mezédes* in this taverna-style restaurant's garden. In winter dine inside by the fire.

3 Erganos Tavern
MAP E6 ▪ Georgiádi 5, Irákleio ▪ 28102 85629 ▪ €

Designed to resemble a Cretan house, Erganos serves local classic dishes such as *sarikópittes* (cheese pies).

4 Remezzo
MAP D6 ▪ El. Venizélou Square 16A, Old Port, Chaniá ▪ 28210 52001 ▪ €

Standing on the water's edge, this landmark restaurant has great *souvláki*, juices and cocktails.

5 To Pigadi Restaurant
MAP E6 ▪ Xanthoudídou Street 31, Réthymno ▪ 28310 27522 ▪ €

Tucked inside a 16th-century building at the foot of a fortress, this eatery specializes in Cretan dishes. Dine inside or in the courtyard.

The courtyard at To Pigadi Restaurant

PRICE CATEGORIES

For a three-course meal for one with half a bottle of wine (or equivalent meal), taxes and extra charges.

€ under €30 €€ €30–€50 €€€ over €50

6 Dionysos
MAP F6 ▪ Ágios Nikólaos lake, Ágios Nikólaos ▪ 28410 25060 ▪ €

This long-established taverna overlooking the lake serves breakfast through to evening meals, and specializes in meat and seafood with Cretan and French influences.

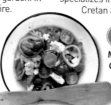

Greek salad with feta cheese

7 Belvedere Restaurant
MAP D6 ▪ Ano Platanias, Chaniá ▪ 28210 60003 ▪ €

Overlooking the sea, Belvedere offers great Cretan food, with *stifádo* (stew) being its signature dish. Greek musicians complete the experience here.

8 Café Greco
MAP E6 ▪ Adélianos Kampos ▪ 28310 73980 ▪ €€

Housed in a modern building with trendy decor, Café Greco is a great place for people-watching, and serves Cretan dishes with produce fresh from the markets, plus a full range of cocktails and good fine wines.

9 Portes
MAP D6 ▪ Portou 48, Chaniá ▪ 28210 76261 ▪ €€

Nestling under the city walls, Portes offers adventurous Greek cuisine that sets it apart from its neighbours.

10 Pacifae
MAP F6 ▪ Ágios Nikólaos lake, Ágios Nikólaos ▪ 28410 24466 ▪ €

Mediterranean dishes with an emphasis on Greek cuisine are prepared by the chefs at this waterside eatery. International and Cretan wines are served.

Places to Shop

(1) Ikaros
MAP E6 ■ Sellia ■ 28320 31271
Set in a beautiful hilltop village, high above Plakias Bay, Ikaros has stylish, unique jewellery, fashioned by silversmith Yannis. Pieces can also be made to the client's own designs or specifications.

(2) Kitamekialo
MAP E6 ■ Arkadíou 184, Réthymno ■ 28310 40611
The designers here combine traditional with modern styles to create playful jewellery for men and women. Products include watches and jewellery made with gold, silver and gems.

(3) Croesus
MAP E6 ■ Arkadíou Street, Réthymno ■ 28102 87924
Designer names such as Cartier, Rolex and Montblanc fill the shelves of men's and women's watches at this shop. The range of contemporary jewellery features gold, silver and gems.

(4) Savoidakis Sweet Shop
MAP E6 ■ Knossou 260, Irákleio ■ 28103 22045
Artfully presented chocolate and *glyká koutalioú* (candied fruit) confectionary are among the tempting displays at this shop.

(5) Kreta Gold
MAP D6 ■ Chálidon Street 71, Chaniá ■ 28210 42623
This jewellery store specializes in local handmade gold and silver necklaces, bracelets and charms, along with religious icons.

(6) Stamatakis Andreas
MAP D6 ■ Selínou 128, Chaniá ■ 28210 95926
One of several branches scattered throughout Chaniá, this bakery serves delicious handmade sweet and savoury pastries, cakes and chocolates. Try its honey-drenched baklava or mini cheese pies.

Clay pots, Creta Ceramics

(7) Creta Ceramics
MAP E6 ■ Thrápsano Pediádos, Irákleio ■ 28910 41717
The age-old craft of creating vases, bowls and *píthoi*-style clay pots is demonstrated at this delightful village workshop and shop.

(8) Mystis Book Store
MAP D6 ■ Chálidon Street 89, Chaniá ■ 28210 92512
Both Greek and international books, many in bilingual editions, along with local music and maps of Crete are available here.

(9) Idols Art
MAP E6 ■ Kato Archanes, Irákleio ■ 28107 52526
This workshop and shop specializes in recreated figurines from Crete's Bronze Age. These handmade items make ideal gifts or souvenirs.

(10) Cicada Gallery
MAP F6 ■ Ágios Nikólaos Road, Eloúnda ■ 28410 41684
With an eclectic range of stone sculptures, plaques and unique jewellery, this is a great place to buy contemporary and traditional Greek gifts of quality.

See map on pp100–101

ᴛᴏᴘ**10** The Dodecanese

Scattered along the coast of Turkey, the Dodecanese is the most southerly group of Greek islands, its pretty whitewashed villages and fine beaches attracting many visitors. The largest islands are Rhodes, Kárpathos and Kos, all of which are popular holiday spots. Many smaller islands and islets

Mosaic, Monastery of St John, Pátmos

make up the rest of the group. The history of the Dodecanese spans several millennia, from ancient times when the Dorians settled here, through the invasion of the Knights Hospitallers in the 14th century, the Ottomans in the 1500s and the Italians in 1941 – all these nations left their mark on the look and culture of the islands.

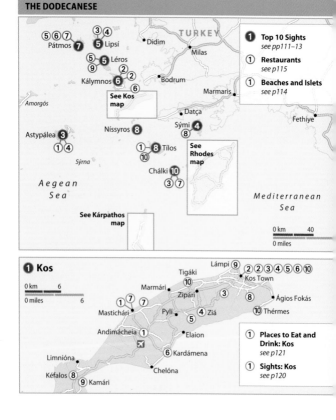

THE DODECANESE

TURKEY

⑤⑥⑦ ③④ Didim
Pátmos ⑦ ⑤ Lipsí · Milas
⑤ ⑤ Léros
⑨ ②②
Kálymnos ⑥ · Bodrum
⑥ Marmaris
See Kos map
Amorgós · Datça
Nissyros ⑧ Sými ④ Fethiye
Astypálea ③ ⑧
①④ ①⑧ Tílos **See Rhodes map**
Sýrna ⑩
Chálki ⑩
Aegean ③⑦
Sea Mediterranean Sea
See Kárpathos map
0 km 40
0 miles

1 Top 10 Sights
see pp111–13

1 Restaurants
see p115

1 Beaches and Islets
see p114

1 Kos

Lámpi ⑨ ②③④⑤⑥⑩
Tigáki ⑩ · Kos Town
0 km 6 Marmári Zipári ③ ⑧ · Ágios Fokás
0 miles 6 ①⑦ ⑦ Pyli ④ Ziá ⑩ · Thérmes
Mastichári ⑤
Andimácheia ① · Elaion
✈
Limniónα ⑥ Kardámena
Chelóna
Kéfalos ⑧
⑨ Kamári

1 Places to Eat and Drink: Kos
see p121

1 Sights: Kos
see p120

1 Kos
MAP G5

The home of Hippocrates, the father of medicine, Kos dates from Neolithic times and was a founding member of the Dorian Hexapolis (see p38) in 700 BC. Life today revolves around its capital, Kos Town, which is dominated by the 15th-century Castle of the Knights. Cafés and chic restaurants sit beside ancient ruins. A marina and a vibrant nightlife characterize the modern town, created after an earthquake destroyed the old town in 1933.

2 Rhodes
MAP G5

Rhodes is famous for being a major stronghold of the Knights of St John who conquered the island in 1306 and stayed for over 200 years. Their legacy can be seen across the island, and especially in Rhodes Old Town (see pp14–15). Along the coast is delightful Líndos, with its narrow cobbled streets, pretty houses and acropolis.

The ancient acropolis, Rhodes

(1)	Places to Eat and Drink: Rhodes *see p117*
(1)	Sights: Rhodes *see p116*

(1)	Places to Eat: Kárpathos *see p119*
(1)	Sights: Kárpathos *see p118*

THE THREE GRACES

In Greek and Roman mythology, the Three Graces were ancient goddesses. They were believed to be the daughters of Zeus, the god of thunder, and his wife Eurynome. The youngest, Aglaea, was the goddess of beauty. Her sisters, Euphrosyne and Thalia, were the goddesses of joy and good cheer.

③ Astypálea

MAP F5 ■ Archaeological Museum: 22430 61500; open Apr–Oct: 8am–8pm Tue–Sat, 9am–3pm Sun & Mon

The small island of Astypálea, which means "old city", was ruled by a Venetian family for several hundred years. They gave it the style it retains to this day. Its capital, Hóra, has many white houses with brightly painted windows and doors that dot the landscape as it rises towards a fortified castle. Finds from the island's prehistoric period are housed in the Archaeological Museum.

④ Sými

MAP G5 ■ Archaeological Museum: 22410 70010; open 8am–2:30pm daily; adm

A popular destination for holidays, Sými is a mountainous island with villages, beaches and bustling harbours. The streets of its capital, Sými Town, and its main port of Gäios are lined with stately mansions from the 19th century, when the island had a thriving boat-building industry. Said to be the birthplace of the ancient Greek goddesses the Three Graces, and taking its name from Syme, Poseidon's wife, Sými has a history spanning millennia. Its museum has some important artifacts.

⑤ Léros and Lipsí

MAP F4

With its pretty valleys and slow pace of life, Léros is a place to unwind. It attracts writers, musicians and artists, who have helped create its identity as a cultural centre. The capital, Agía Marina, ascends from the harbourside through alleyways of Neo-Classical buildings and white cottages to a hilltop dominated by the remains of a Knights' castle. Lakkí, the island's main port, is one of the largest natural harbours in the Mediterranean, from where boats go to Lipsí, an enchanting island full of colourful cottages.

⑥ Kálymnos

MAP F4 ■ Archaeological Museum, Kálymnos: 22430 23113; open Mar–Nov: 8:30am–3pm Tue–Sun; adm

Remains of ancient settlements testify to Kálymnos's long history. Locally found Neolithic, Minoan and Mycenaean figurines are on display at its Archaeological Museum. Its capital is Póthia, a picturesque town where brightly coloured houses of local spongefishers line the deep harbour in amphitheatre-fashion. A church dedicated to Christ the Saviour, noted for its frescoes and iconostasis, is Póthia's principal landmark. Most of the rest of the isle is mountainous.

⑦ Pátmos

MAP F4

Known as the Jerusalem of the Aegean, Pátmos is both a holiday island and a place of pilgrimage. At its heart is the UNESCO World Heritage Site, Horá, which boasts over 40 Byzantine monasteries and chapels, including the significant Monastery of St John *(see pp16–17)*.

Pastel-coloured houses, Sými

View down through the valley, Tílos

8 Níssyros and Tílos
MAP G5

A volcanic island, Níssyros is known for the unique species of flora and fauna that grow in its mineral-rich soil. It is possible to visit the crater of the extinct volcano at Polyvótes. Villages and towns lie in the lower coastal plains. Nearby, the rugged island of Tílos is on a bird migration path and a popular birdwatchers' spot.

9 Kárpathos
MAP X5

The group's second-largest island, Kárpathos is characterized by its indented coastline, sandy beaches and high mountains. The highest is Kalí Límni, at 1,214 m (3,983 ft). The capital, Pigádia, is cosmopolitan. The island has bustling resorts, such as Ammopí and Lefkós. Thriving traditional rural villages, such as Menetés, where artifacts have been discovered from the Middle Ages, and Óthos, famous for its traditional Kárpathian houses, are worth a visit.

10 Chálki
MAP G5

Reached via a boat ride from Rhodes, Chálki is a small island that revolves around its only community, Emborió, with 200 or so inhabitants. It boasts the Dodecanese's highest bell tower, on the golden-stone Ágios Nikólaos church. Visitors can enjoy its rural feel, its beaches, such as Ftenágia (see p51) and Trachiá, and the deserted Chorió village, with its Knights' castle.

DAY TRIP FROM RHODES TO CHÁLKI

▶ MORNING

Although the Dodecanese islands are strung out and the only way to see all of them is to sail or take ferries over several days, it is possible to take a day trip from one island to another. A popular route is from **Rhodes Town** to **Emborió** on **Chálki**. Boats also leave from Skála Kameírou (see p116) on Rhodes to Chálki.

After a good breakfast, head for Mandráki harbour in Rhodes Town. You will need to get there early as most boats and hydrofoils to Chálki leave at about 9am. It is a good idea to check the timings a day before at a travel agency. Hop on board for the 2-hour sail that takes you along the coastline, past the islet of Alímnia to the gorgeous harbour of **Emborió**. Refreshments are usually available on board. Upon disembarking, take time to walk along the harbourside and enjoy a delicious lunch at one of the tavernas that line the harbour.

AFTERNOON

Take a taxi to see the imposing **Crusader Castle** that sits high above the harbour. It is well worth the trip and the view across to Rhodes is rewarding in itself. You can also visit some of the island's excellent churches and monasteries, such as the **Moní Agíou Ioánnou Prodrómou**, which has breathtaking icons.

Refresh yourself with a drink or snack at Ftenagia Restaurant (see p115) near Ftenágia beach (see p51), before making your way back to the harbour in time for the return crossing.

See map on pp110–11

Beaches and Islets

1 Ágios Antónis Beach, Tílos

MAP G5 ▪ Ágios Antónis village

High and rugged mountains plunge to meet this sandy beach and its crystal-clear blue bay. The view is dramatic.

2 Arginónta Beach, Kálymnos

MAP F4 ▪ West coast near Masoúri

A forested hillside descends almost to the water's edge here, gently giving way to the sand and pebble Arginónta beach. Tavernas are located in a village nearby.

3 Katsadiás Beach, Lipsí

MAP F4 ▪ Katsadiás

A series of sandy coves lead to the beautifully wide sand and shingle Katsadiás beach. It is a popular anchorage for yachts due to its seclusion, shelter and shallow water.

4 Tría Marmária Beach, Astypálea

MAP F5 ▪ Astypálea Town

These three beaches are collectively known by a single name. Each is horseshoe-shaped and secluded, enclosed by a rugged landscape and deep bays of azure water.

5 Psilí Ámmos Beach, Pátmos

MAP F4 ▪ Southwest coast

A long stretch of golden sand, dotted with trees and sandwiched between turquoise sea and mountains, this is the best beach on the island. Accessibility is difficult *(see p50)*.

6 Psérimos Islet

MAP F4

This tiny islet has one of the best beaches in the Dodecanese. It is long, with soft sand and shallow waters, and is popular with day trippers.

7 Póntamos Beach, Chálki

MAP G5 ▪ Emborió

Chálki has several beautiful beaches but Póntamos is the only one you do not need a boat to reach. Golden sand and shallow water make this beach ideal for children.

8 Faliráki Beach, Rhodes

MAP V4 ▪ East coast

More famous for the nightlife that lines the waterfront, this beach is exceptionally beautiful. About 4 km (2.5 miles) long, it has golden sands and safe swimming waters. Located nearby is the quieter Kathará beach.

9 Finíki Beach, Kárpathos

MAP X6 ▪ Finíki village

Lying within a horseshoe-shaped bay and set among hills dotted with the whitewashed, red-roofed houses of Finíki village, this sandy beach is one of the prettiest on the island.

10 Thérmes Beach, Kos

MAP Y2 ▪ East coast

With its natural hot springs and an underground spa, Thérmes beach is popular with locals as well as visitors. Accessibility is good, and car parking facilities and tavernas plentiful.

Mountains shielding Psilí Ámmos

Restaurants

PRICE CATEGORIES

For a three-course meal for one with half
a bottle of wine (or equivalent meal),
taxes and extra charges.

€ under €30   €€ €30–50   €€€ over €50

1 Astifagia, Astypálea
MAP F5 ■ Paralia Schoinonta
■ 22430 64004 ■ €€

Serving fish and seafood, and a menu
of authentic Greek meat dishes using
organic produce, this informal eatery
is near the sea. Lobster and prawns
in *ouzo* are two of its specialities.

2 Medusa by the Sea, Kálymnos
MAP F4 ■ Harbour, Rína ■ 69727
92482 ■ €€

Full of character, this harbour
restaurant with stone walls and a
pretty terrace serves fresh fish and
homemade dishes. The owners grow
almost all the produce themselves.

3 Ftenagia Restaurant, Chálki
MAP G5 ■ Ftenágia beach ■ 22460
45384 ■ €€

Housed in a rustic building, this
elegant restaurant has an excellent
Greek menu. Roasted octopus, Chálki
shrimp and *mezédes*, along with local
wine and beers, tempt the tastebuds.

4 Pefko, Lipsí
MAP F4 ■ Harbour, Lipsí Town
■ 22470 41404 ■ €

Near the Ágios Ioánnis church, Pefko
offers grilled classics, like *souvláki*
and fresh octopus, accompanied by
local wines, on its pretty terrace
overlooking the harbour.

5 Taverna Paradisos, Léros
MAP F4 ■ Vromolithos ■ 22470
24632 ■ €

In an idyllic spot right on the beach,
this family-run taverna serves tasty
food cooked to authentic homemade
recipes. The service is excellent.

6 Trehantiri Taverna
MAP F4 ■ Skála ■ 22470
34080 ■ €

Located in a small street, this
modest taverna serves generous
portions of authentic Greek seafood
dishes and local specialities.

Table setting, Benetos Restaurant

7 Benetos Restaurant, Pátmos
MAP F4 ■ Sapsíla, near Skála ■ 22470
33089 ■ €€

Dine on fish, served with herbs and
lemon, on the terrace of this super
restaurant overlooking the sea.

8 Taverna to Spitiko, Sými
MAP G5 ■ Gialós ■ 22460
72452 ■ €€

This taverna is famed for its platters
of local shrimps, squid and mussels,
and for its idyllic location on the
seafront overlooking colourful Neo-
Classical houses across the bay.

9 Zorbas, Léros
MAP F4 ■ Pantelí beach,
Pantelí ■ 22470 22027 ■ €€

This landmark restaurant, located
next to a pretty beach, serves Lerian
dishes, such as its own take on
moussakás with local herbs.

10 Armenon, Tílos
MAP G5 ■ Livadia ■ 22460
44134 ■ €€

Dishes made with local produce are
served at this taverna with unrivalled
views across Livadia Bay. Olive oil
and honey is made in-house.

See map on pp110–11

Sights: Rhodes

1 **Asklipeió**
MAP U5 ■ Inland from Líndos
Home to fine 17th-century frescoes, the Dormition of the Virgin church has put this pretty village on the map.

2 **Ancient Ialyssos**
MAP V4 ■ Near Triánda
This mountainside city was part of Rhodes' ancient city-state and retained its importance during Byzantine, Knights and Ottoman times. It is the site of Moní Filerímou.

Archaeological site, Kámeiros

3 **Ancient Kámeiros**
MAP U4 ■ Kámeiros, near Kalavárda
This fascinating site from the 5th century BC is one of the finest examples of Doric city planning. It is possible to see remnants of a temple, dwellings and public baths.

4 **Líndos Acropolis**
MAP U5 ■ Líndos
At this archaeological site, the ancient acropolis and a 3rd-century-BC temple dedicated to the goddess Athena are the most notable remains of a great Doric city-state formed by Líndos, Ialyssos and Kámeiros.

5 **Monólithos**
MAP T5 ■ Near Foúrnoi
This quiet village is most famous for a 15th-century fortress built on a sheer craggy rock by Grand Master d'Aubusson of the Knights Hospitallers. There is also a lovely beach nearby, Foúrnoi.

6 **Petaloúdes**
MAP U4 ■ Petaloúdes
Known as the Valley of Butterflies, this area has numerous trees that secrete a vanilla-scented gum, used for incense. It attracts thousands of butterflies and moths.

7 **Skála Kameírou**
MAP T5
This once major ancient city port has one of the most impressive castle ruins on the island – the 16th-century Castle of Kritinía.

8 **Profítis Ilías**
MAP U5 ■ Inland from Kalavárda
The densely forested Profítis Ilías mountain was a natural park under Italian rule. Today it is famous for its nature trails and walks.

9 **Palace of the Grand Masters**
MAP V4 ■ Rhodes
The palace is built at the highest point of the medieval city and is its most prominent landmark *(see p14)*.

10 **Moní Filerímou**
MAP V4 ■ Near Triánda
A place of worship for over 2,000 years, the Moní Filerímou is famous for the Our Lady of Filérimos complex of four chapels.

Hillside chapel at Moní Filerímou

Places to Eat and Drink: Rhodes

1 Ambrosia
MAP G5 ▪ Líndos Village, Líndos ▪ 22440 31804 ▪ €€

This elegant restaurant with subtle lighting and crisp white tablecloths serves imaginatively plated dishes from around the islands, with a focus on fish and seafood.

2 Philosophia Beach Taverna
MAP U4 ▪ Péfkoi ▪ 22440 48044 ▪ €€

In a prime spot overlooking a beach, this taverna offers the chance to dine alfresco on homemade Greek dishes prepared using organic produce from its own gardens. It is known for its delicious desserts.

3 Kalypso Roof Garden Restaurant
MAP U4 ▪ Líndos ▪ 22440 32135 ▪ €€

With Líndian ceramics and nautical memorabilia adorning its walls, this delightful restaurant in a 17th-century waterfront house serves great fish.

4 Socratous Garden
MAP G5 ▪ Rhodes Town ▪ 22410 76955 ▪ €

Stop at this lovely café for a light lunch or cocktail on the outdoor terrace. At night, enjoy a hearty meal, drinks and DJ spun tunes (see p61).

5 Alexis 4 Seasons
MAP V4 ▪ Aristotélous Street 33, Rhodes Old Town ▪ 22410 70522 ▪ €€€

The acclaimed Alexis is housed in a historic building and serves fabulous fish and classic Greek dishes. A roof garden offers great views (see p65).

6 Tamam
MAP V4 ▪ Leontos Georgiou 1, Rhodes Town ▪ 22410 73522 ▪ €€

This small and quaint taverna serves an extensive menu of traditional Greek food, all cooked by Maria, the owner's daughter. The house wines are good quality and value for money.

PRICE CATEGORIES

For a three-course meal for one with half a bottle of wine (or equivalent meal), taxes and extra charges.

€ under €30   €€ €30–€50   €€€ over €50

7 Zig Zag Bar
MAP V4 ▪ Aristotélous Street 33, Rhodes Old Town ▪ €€

A bustling interior and broad terrace make this cocktail bar a hit. A good selection of cocktails and other drinks are on offer (see p60).

8 Romeo
MAP V4 ▪ Corner Menekléous and Sokrátous Street, Rhodes Old Town ▪ 22410 25186 ▪ €€

An array of Greek dishes will tempt the palate at this taverna, located in an early 16th-century house.

Romeo, buzzing with customers

9 Epta Piges Taverna
MAP U5 ▪ Archángelos ▪ 22410 56417 ▪ €

Epta Piges, meaning Seven Springs, is an alfresco taverna nestled under huge pine and plane trees. The menu is classic Greek, with speciality homemade *moussakás*.

10 Restaurant Wonder
MAP V4 ▪ El Venizélou Street 16–18, Rhodes Town ▪ 22410 39805 ▪ €€€

This elegant restaurant, set in a mansion, serves international cuisine with a Swedish and Asian twist.

See map on pp110–11

Sights: Kárpathos

1 **Pigádia (Kárpathos Town)**
MAP Y6

Traditional tavernas line the pretty harbourside of this modern capital, which has grown from an ordinary port town into a centre of finance and commerce. The adjacent Vróndis bay boasts a lovely beach.

2 **Menetés**
MAP X6

This charming village's main street is lined with pastel-coloured Neo-Classical houses and interspersed by tiny alleyways of steep steps.

3 **Arkása**
MAP X6

Arkása is a bustling resort, but it is also the site of the ancient city of Arkesia. You can see ruins of the 4th-century church of Agía Anastasía, where several important archaeo-logical finds were discovered.

4 **Apéri**
MAP X5

A village of elegant whitewashed mansions and squares that sit amphitheatre-fashion in the foothills of Mount Kalí Límni, Apéri was the capital of Kárpathos until 1892.

5 **Lefkós**
MAP X5

This pretty village, a top holiday spot, has white sandy beaches nestled in deep coves with crystal-clear water and a backdrop of pine trees.

6 **Diafáni**
MAP Y4

A picturesque village that revolves around its intermittently busy harbour, Diafáni is popular with day-trippers that come here for the good sandy beaches and family tavernas.

7 **Ammopí**
MAP Y6

The Ammopí resort is best known for its beaches, which are among the best on the island. Mikrí Ammopí occupies a crescent of pristine golden sand. Megáli Ammopí and Votsalákia are also very pleasant.

8 **Surrounding Islands**

To the north is Saría, said to be the site of the ancient Níssyros city, while to the south is unspoiled Kássos. Other islets include Platý, Armathiá, Kouríka and Mákra.

9 **Ólympos**
MAP X4

Climbing the hillside, with its working windmills, painted stone houses and locals in traditional dress speaking a Dorian dialect, Ólympos is like a living museum.

10 **Óthos**
MAP X5

A pretty village of traditional Karpathian houses, Óthos is the island's highest community at 450 m (1,500 ft). It is known for its red wine and festivals.

Lefkós bay and its beach

Places to Eat: Kárpathos

1 To Ellinikon
MAP Y6 ■ Pigádia ■ 22450 23932 ■ €

Along with seafood and grilled meat fresh from the charcoal, this small eatery specializes in salads. Try the Venus salad, served with the local *haloúmi* cheese. Dine inside or out.

Traditional Manolis Taverna

2 Manolis Taverna
MAP X6 ■ Menetés ■ 22450 81103 ■ €€

Housed in one of Menetés' traditional little houses, this taverna has a menu of wholesome dishes, such as *souvláki* and chicken char-grilled with herbs.

3 Potali Bay
MAP X5 ■ Potáli Bay ■ 22450 71221 ■ €

Fish is high on the menu at this place right on the beach. Choose from octopus, shrimps and lobster, served with added extras like pumpkin cakes.

4 Acropolis
MAP Y6 ■ Pigádia ■ 22450 23278 ■ €€

Right by the harbour, Acropolis is popular with locals and serves wholesome bistro-style meat and fish dishes with fine wines.

5 Orea Karpathos
MAP Y6 ■ Pigádia ■ 22450 22501 ■ €

This attractive eatery serves *mezédes* with lots of local dishes plus a great selection of fish and grills.

PRICE CATEGORIES

For a three-course meal for one with half a bottle of wine (or equivalent meal), taxes and extra charges.

€ under €30   €€ €30–€50   €€€ over €50

6 Dolphin Taverna
MAP X6 ■ Finíki ■ 22450 61060 ■ €

Overlooking Finíki beach and the working fishing harbour, this taverna serves authentic Karpathian delicacies. The *mezédes* and fish cooked with local herbs and cheese are particularly good.

7 Mesogeios Restaurant
MAP Y6 ■ Pigádia ■ 22450 23274 ■ €€

Diners sit at bright blue tables to enjoy exquisite local dishes from a tapas-style menu at this lively restaurant standing right on the waterfront *(see p65)*.

8 To Perasma
MAP Y6 ■ Pigádia ■ 69786 22904 ■ €

Located just across the road from the ruins of the Agía Fotini Chapel, this family restaurant serves delectable Mediterranean cuisine at reasonable prices.

9 Milos Taverna
MAP X4 ■ Ólympos ■ 22450 51333 ■ €

This taverna has been built around a working windmill with a brick oven as its centrepiece. The menu features local dishes, such as savoury *píttes* (pies), which are prepared using organic produce.

10 Glaros Restaurant
MAP X6 ■ Ágios Nikólaos beach, Arkása ■ 22450 61015 ■ €

An alfresco restaurant that overlooks the beach, the Glaros specializes in Karpathian and Greek dishes such as stuffed calamari. It has a special kids' menu and delicious desserts.

See map on pp110–11

Sights: Kos

Windmill, Andimácheia

1 Andimácheia
MAP X2

Particularly known for its windmills, this traditional village is dominated by the imposing 14th-century Castle of Andimácheia, which was built by the Knights Hospitallers as a prison.

2 Kos Town
MAP Y1

The 16th-century Castle of the Knights and ancient remains such as the 3rd-century-BC agora dominate the capital city, Kos Town. Restaurants along palm-lined avenues give it a cosmopolitan feel.

3 Asklepieíon
MAP Y1

Once a centre of healing, Asklepieíon is one of Greece's most important historic sites. The major highlights include remains of the 3rd-century Temple of Apollo, Hippocrates' garden and the Roman baths.

4 Ziá
MAP X2

Preserved as a model village, Ziá has a Byzantine church and stone houses lining cobbled pathways. It is one of the Asfendioú Villages (see p53).

5 Paleó Pýli
MAP X2

Spectacularly perched on a clifftop, with its walls built into the rock, this abandoned fortified town is famous for its Byzantine churches.

6 Kardámena
MAP X2

One of Greece's liveliest resorts, Kardámena, a former fishing village, offers watersports, boat trips, nightclubs, bars and tavernas.

7 Mastichári
MAP X2

Boasting some of the best beaches on the west coast, this pretty fishing harbour is popular with locals. Its tavernas serve fabulous fish.

8 Kéfalos
MAP W2

An elevated village of whitewashed houses and cobbled paths descending to the beach, Kéfalos is home to the ruined Castle of the Knights.

9 Kamári
MAP W2

With its volcanic backdrop and soft, sandy beach, this village is a popular holiday spot. Resorts and restaurants line its waterfront.

A flamingo gathering, Tigáki

10 Tigáki
MAP X1

Tigáki is known for its abundance of wetland bird species, which flock to its salt lakes. The village also has sandy beaches and shallow waters.

See map on pp110–11

Places to Eat and Drink: Kos

1 O Makis
MAP X2 ▪ Mastichári ▪ 22420 59061 ▪ €€

Specializing in Greek dishes, such as *souvláki* served straight from the grill with parsley and lemon, this super little taverna near the harbour is an informal place to dine.

2 Avanti Restaurant
MAP Y1 ▪ Vasileíou Georgíou, Kos Town ▪ 22420 20040 ▪ €€

This cheerful waterfront café and pizzeria has an extensive Italian-themed menu featuring savoury homemade crêpes and pizzas cooked in its wood-fired oven.

3 Platanos
MAP Y1 ▪ Plateía Platanoú, Kos Town ▪ 22420 28991 ▪ €€

This large restaurant is right next to the Castle of the Knights and the plane tree here is said to have been planted by Hippocrates. The menu is Greek and French.

4 Kalymnos Restaurant
MAP Y1 ▪ Kos Town ▪ 22420 48540 ▪ €€

The family-run Kalymnos is a pretty taverna minutes from Kos Town's main beach. It specializes in home cooking, with Greek recipes handed down through the generations.

5 Restaurant Museum
MAP Y1 ▪ Ríga Feréou Street 2, Kos Town ▪ 22420 20999 ▪ €€

Located near the Archaeological Museum, this attractive eatery serves international fare, often straight from the coals, great seafood and vegetarian dishes, homemade desserts and local wines. Dine inside or alfresco.

6 Ristorante Otto e Mezzo
MAP Y1 ▪ Apéllou Street 21, Kos Town ▪ 22420 20069 ▪ €€

Dine on dishes inspired by the regions of Italy. Terraces overlook the Old Town and the gardens.

7 Tam Tam
MAP X2 ▪ Mastichári ▪ 69444 37027 ▪ €

Enjoy grills and cocktails at this informal beach bar under rustic straw parasols and palm trees.

Great views from Lofaki Restaurant

8 Lofaki Restaurant
MAP Y1 ▪ Ágios Nektarios, Kos ▪ 22420 21982 ▪ €€

One of Kos's premier eateries, Lofaki serves artfully presented Greek dishes with local and inter-national wines.

9 Delon Pub, Kos
MAP Y1 ▪ Lampi ▪ 22420 22824 ▪ €

Operating for over 35 years, this family-run bar offers a great atmos-phere and superb beer *(see p60)*.

10 Restaurant Jumbo Style
MAP Y1 ▪ Plateía Agías Paraskevís, Kos Town ▪ 22420 24780 ▪ €€

Housed under the canopy of the covered market, this popular grill house is informal by day and lively in the evening. *Mezédes* are a speciality.

Following pages Archaeological remains on the island of Delos

🔟 The Northeast Aegean Islands

A collage of ancient ruins, fabulous beaches, quaint villages and lush countryside, this archipelago is one of the least touched by modern-day tourism. The group comprises the larger islands of Límnos, Lésvos, Híos, Sámos and Ikaría, and also includes Thássos, inhabited since the Stone Age, and undeveloped Samothráki, both up near the Macedonian coast. The islands share a landscape of rugged mountains, indented coastlines and glorious beaches. Tiny islets dot the surrounding waters, including Foúrni, Ágios Efstrátios, Psará and the wildlife haven of Andípsara.

Archaeological artifact found on Sámos

THE NORTHEAST AEGEAN ISLANDS

1	**Top 10 Sights**	see pp125–7
1	**Places to Eat: Lésvos**	see p133
1	**Sights: Lésvos**	see p132
1	**Places to Eat: Híos**	see p131
1	**Sights: Híos**	see p130
1	**Restaurants**	see p129
1	**Islands, Bays and Beaches**	see p128

Windmills dotting the landscape, Híos

1 Híos
MAP K5

A sizable island, Híos has a complex history, including a spell of being one of the wealthiest islands in the Mediterranean through its trade in gum mastic. The *mastichochória* (mastic villages) where the industry's workers lived can still be visited today *(see p130)*. They lie to the south of the island's relatively modern capital, Híos. Nearby is the Néa Moní, an 11th-century monastery *(see pp20–21)*.

2 Sámos
MAP F4

Sámos is a green island famous for being a major maritime power in antiquity and for the important archaeological sites of Pythagóreio and Heraion *(see pp24–5)*. Its capital is Vathý, an attractive harbour town with cobbled lanes lined with red-roofed Neo-Classical mansions. Beyond are lush mountains covered with valleys of vineyards. Vathý and its surrounding villages are famous for their golden sweet Muscat wine.

3 Límnos
MAP E1

In Greek mythology, Límnos was where Hephaestus, the god of metal-working, landed after being hurled out of Olympus (home of the gods) by his angry father Zeus. A volcanic island, Límnos, or Lemnos, is largely flat with vineyards planted on lava-rich soil sloping towards wide, sandy beaches. Its capital is Mýrina, a town of Neo-Classical and Ottoman buildings and cobbled streets, dominated by its imposing Venetian *kástro* (castle). The island is known for its herbal honey.

Lésvos

Míthymna
Sykaminiá
Pétra
Mantamádos
Ándissa
Agía Paraskeví
Sigrí
Kallóní
Mesótopos
Pigí
Thermís
Polìchnitos
Agiásos
Mytilíni Town
Vaterá
Plomári

km 15
miles 15

Híos

Agio Gála
Kardámyla
Inoússes
Potamiá
Volissós
Langáda
Vrontádos
Avgónyma
Híos Town
Véssa
Kámbos
Mestá
Armólia
Olýmbi
Pýrgi
Emboreiós

km 10
miles 10

Remains of the castle, Mýrina, Límnos

4 Archaeological Museum Pythagóreio, Sámos

MAP F4 ▪ Pythagóreio ▪ 22730 62813 ▪ Open 8am–3pm Tue–Sun ▪ adm

This modern museum exhibits archaeological finds from the ancient city of Sámos in chronological and thematic order from the 9th century BC to the 7th century AD *(see p42)*.

5 Ikaría
MAP F4

Ikaría is named after Icarus, the foolhardy son of ancient craftsman Daedalus. According to legend, Ikaros fell to his death after flying too close to the sun wearing artificial wings. The island has been inhabited since Neolithic times, and was part of the Genoese Aegean Empire and later the Ottoman Empire until its independence in 1912. Once a favourite haunt of pirates, which is evident in the defensive layout of its villages, Ikaría has rich soil which produces fine wines.

View over the rooftops, Ikaría

6 Psará and Andípsara
MAP E3 ▪ Off Híos

Psará is best known for its heroic Freedom or Death flag, to which Psariots have been faithful in the face of battle. The most celebrated stance was in 1824 when, faced with the Ottoman invasion, residents blew themselves and their invaders up with gunpowder. Psará's neighbour is the uninhabited Andípsara island, an important environmental and wildlife sanctuary. Many interesting birds roost here, including Eleonora's falcon *(see p55)*.

Ruins of the Sanctuary of the Great Gods on the island of Samothráki

7 Samothráki
MAP E1

With the 1,600-m- (5,250-ft-) high Mount Fengári, hot thermal springs, olives groves, forests of oak and chestnut and two spectacular waterfalls, Samothráki has one of the most dramatic landscapes of all the eastern Aegean islands. At its heart is the capital Hóra, an elegant town of squares and cobbled streets. Nearby is the ancient capital of the island, Palaiópolis, and the remains of the Sanctuary of the Great Gods.

8 Foúrni
MAP F4

It is here where most Foúrni residents live and work, and where visitors can find small tavernas and excellent shops. An island of cliffs, deep bays and long stretches of sandy beach that once made it a pirates' paradise, this is a popular day-trip destination from the nearby islands of Sámos, Pátmos and Ikaría.

9 Lésvos
MAP Q1

The third-largest Greek island, Lésvos revolves around its capital, the elegant city of Mytilíni. The town of Agiásos is considered the island's most picturesque and is famous for

Eleonora's falcon, Andípsara

Sámos (40km)
Chrisomiliá
Psilí Ámmos
Ikaría (20km)
Foúrni Town
Foúrni
Kambí

▶ MORNING

Foúrni can be reached easily by ferry from Vathý harbour or Pátmos harbour on Sámos (see p125) and Ágios Kírykos on Ikaría. Foúrni Ferries and Greek Ferries are two of the companies serving the island. Departure and return times, as well as prices, vary, so do check in advance. After a hearty breakfast, head to the port of your choice. Most ferries leave between 9am and 10am and crossing times are short (under 20 minutes from Ikaría). Ferries pull into Foúrni's harbour, a lively place of working fishermen who mingle with visitors and local families going about their daily business. Promenade cafés offer the chance of a mid-morning coffee and snack.

From here, head into Foúrni Town, which centres around its main square, linked with the harbour by a single tree-lined street. Take time to admire the traditional Eastern Aegean architecture and shop along the agorá, known locally as the "shopping mile". Be sure to stop for lunch in one of the excellent fish tavernas for which Foúrni is famous.

AFTERNOON

After lunch, take a short walk to Kambí, famous for its windmills, and Psilí Ámmos, where there are some fine beaches and bays to explore. Alternatively, head north by taxi or boat to the small, sleepy village of Chrisomiliá, where life has changed little in many decades, before heading back to the harbour for your return crossing.

owning an icon believed to have been painted by St Luke. Villages of interest include Ypsiloú, near a fossilized forest and an extinct volcano. Lésvos, or Lesbos as it is often referred to, is known for being the birthplace of Sappho, the ancient female poet who wrote erotic poems to other women. Lésvos has become the focal point of the refugee crisis in Greece as the number of migrants arriving on its shores has increased dramatically. Lésvos's residents have been nominated for a Nobel Peace Prize for their help, but worries persist over the long-term effects to its tourism.

⑩ Thássos
MAP E1

This almost circular island has remains from the Bronze Age that suggest it had strong links with the Cyclades. In the 7th century BC, settlers from Páros colonized parts of Thássos and, with its natural gold and marble resources, the island has known wealth and power.

SANCTUARY OF THE GREAT GODS, SAMOTHRÁKI

This sanctuary dates from Pan-Hellenic times and was the most important religious site for cult worship in ancient Aeolia, Thrace and Macedonia. Followers included the Spartan leader Lysander. The sanctuary grew in the Hellenistic era. Today, its remains include a cult initiation room and an amphitheatre.

See map on pp124–5 ←

Islands, Bays and Beaches

The beautiful Kokkári beach stretching around a curved bay

1 Kokkári Beach, Sámos
MAP F4 ■ North coast

Catching strong winds, Kokkári is one of the islands' best beaches for windsurfing. The pebbled beach lies next to a tourist resort, which has a variety of amenities.

2 Elínda Beach, Híos
MAP K5 ■ West coast, near Anávatos

This is a wonderful secluded beach of fine, golden sand and crystal-clear water within a deep bay. Surrounding hills protect it from strong winds.

3 Mávros Gialós, Híos
MAP K6 ■ South coast, near Emboreiós

Also known as Mávra Vólia, this beach lines a bay of turquoise waters. It has black pebbles, created by the lava of a nearby volcano and worn smooth by the sea over time.

4 Mykáli Beach, Sámos
MAP F4 ■ Southeast coast, near Psilí Ámmos

Popular with locals, this white-pebble beach is one of the longest along this coastline. Mykáli beach is protected from the wind and has no amenities.

5 Pachia Ammos, Samothráki
MAP E1 ■ South coast, near Kamariotissa

This sandy cove offers a beach bar and a taverna to the south, and secluded swimming at the northern end.

6 Alykí Bay, Thássos
MAP E1 ■ Southeast coast

Widely regarded as one of the most beautiful harbours in the Greek Islands, this richly forested bay of pine and olive trees has two sandy beaches, clear waters and unusual rock formations.

7 Inoússes Island
MAP L4 ■ East of Híos

Daily life on this charming island centres around its harbour, where a museum tells the story of its maritime history. The landscape is dotted with olive groves and stone villages.

8 Nas, Ikaría
MAP F4 ■ West of Arménistis

This beach lies in a picturesque cove at the union of the Chalaras river and the Aegean Sea. Look out for the ruins of a temple dedicated to Artemis.

9 Karfás Beach, Híos
MAP L5 ■ East coast, near Híos Town

The liveliest tourist beach on the island, Karfás has fine sand, gentle waters and amenities that include sunbeds, umbrellas and nearby bars and cafés. Watersports are available.

10 Ánaxos Beach, Lésvos
MAP R1 ■ Pétra

Looking across the bay towards Mólyvos, this lively beach is long, sandy and lined with bars and tavernas. There are resorts nearby, making it popular with tourists.

See map on pp124–5

Restaurants

1 Irodion Garden, Sámos
MAP F4 ▪ Aristarxou 34
▪ 22730 61642 ▪ €

Housed in a stylish Neo-Classical villa, this restaurant has several terraces for alfresco dining. There's a broad menu with some delicious specials, such as spare ribs.

2 Karnagio Taverna, Thássos
MAP E1 ▪ Seafront Liménas
▪ 25930 22006 ▪ €€

This seafront *ouzerí* has tables set out on the beach and a roof terrace with great views. Soft music and Greek food help make this a memorable place.

3 Taverna Steki, Thássos
MAP E1 ▪ Potamiá
▪ 25930 61009 ▪ €

Enjoy international as well as local Greek dishes made by the owner, Katerina, at this taverna. Most vegetables come straight from her garden.

4 O Glaros, Límnos
MAP E1 ▪ Harbourside, Mýrina Town ▪ 22540 22220 ▪ €

Known as much for its great views of the *kástro* (castle) as its super menu, this is a great place to enjoy fish. Try the lobster with spaghetti.

5 Anna's Restaurant, Ikaría
MAP F4 ▪ Nas ▪ 06932 149155 ▪ €

Overlooking the sea, this small fish restaurant has a good reputation for its imaginative menu. Try the lobster with lemon and stuffed *kalamári*.

6 Klimataria, Samothráki
MAP E1 ▪ Seafront, Hóra
▪ 25510 41535 ▪ €

Famous for its speciality dish *gianiótiko* (baked pork with potato and eggs), this restaurant has a traditional feel and Greek music.

> **PRICE CATEGORIES**
> For a three-course meal for one with half a bottle of wine (or equivalent meal), taxes and extra charges.
>
> € under €30   €€ €30–€50   €€€ over €50

7 Marina Restaurant, Sámos
MAP F4 ▪ Potámi, Kokkári ▪ 22730 92692 ▪ €

Enjoy dishes made with local produce at this popular restaurant. Dine on the terrace overlooking the gardens to benefit from the sea views.

8 Karydies, Samothráki
MAP E1 ▪ Ano Meria 21 ▪ 25510 98266 ▪ €

Delicious dishes such as *kléftiko* (lamb) with bread, are cooked in traditional wood ovens at this popular taverna.

A sumptious serving of kléftiko at Karydies

9 Poseidon, Sámos
MAP F4 ▪ Harbourside, Kokkári ▪ 22730 92384 ▪ €

With tables set on a terrace right by the waterside and a menu of freshly caught fish, this is one of the most popular restaurants on the island.

10 Zorbas, Thássos
MAP E1 ▪ Skála Prínou ▪ 25931 12356 ▪ €

Near the village centre and the beach, this restaurant serves a wide range of grills, local dishes like *afélia* (pork), *souvláki* and desserts.

Traditional Greek decor, Zorbas

Sights: Híos

1 Mastic Villages
MAP K6

These 20 or so fortified settlements, including Mestá, Pýrgi and Olýmbi, are called *mastihohória* (mastic villages) after the lucrative mastic gum industry they once supported.

A narrow stone street, Olýmbi

2 Olýmbi
MAP K6

With its fortress-like layout – whereby the whole village is contained within a wall, the only entrance being the Kato Porta watchtower – Olýmbi is a fine medieval monument.

3 Pýrgi
MAP K6

Named after a medieval tower that stands here, the village of Pýrgi is best known for its painted houses. Façades are decorated with the grey and white geometric pattern known as *xystá*.

4 Mestá
MAP K6

A fine example of the defensive architecture that characterizes the mastic villages, Mestá's outer stone buildings join to create a wall. Its castle and churches are also notable.

5 Volissós
MAP K4

This picturesque village, boasting restored stone houses arranged around a mountain, is dominated by a ruined Byzantine castle.

6 Kámbos
MAP L6

Wealthy merchants and nobles once had summer mansions in Kámbos, before Híos' destruction in 1822. Citrus groves surround this area.

7 Vrontádos
MAP L5

Known for its landmark windmills, this village hugs the shore and fishing boats line its quayside. Sights include the Moní Agíou Stefánou monastery.

8 Híos Town
MAP L5

With a history that has seen prosperity under the Genoese, a massacre by the Ottomans and huge earthquake destruction, Híos is now an attractive and modern capital.

9 Avgónyma
MAP K5

One of the most pretty hillside villages on the island, Avgónyma has elegantly restored houses and is home to many Greek-Americans.

10 Moní Moúndon
MAP K4

This 16th-century monastery is known for its well-preserved murals. The monastery celebrates St John the Baptist on August 29 and is open to the public from the early morning.

Mural, Moní Moúndon

Places to Eat: Híos

1 To Kechrimpari
MAP L5 ▪ Ágios Anargiron 7, Híos Town ▪ 22710 27541 ▪ €

Fish or meat mezédes, which start with dips and conclude with dessert and an ouzo, are the speciality at this cosy restaurant. The dining area has traditional decor with stone walls adorned with family pictures.

2 Mavrokordatiko Restaurant
MAP L5 ▪ 1 Mitaráki, Kámbos ▪ 22710 32900 ▪ €

Housed in a beautifully restored 18th-century stone building, this restaurant regularly hosts special events, such as Greek nights. Dine on Greek and international dishes in the courtyard or inside.

3 Pýrgos Restaurant
MAP K5 ▪ Avgónyma ▪ 22710 42175 ▪ €

One of the oldest tavernas in Avgónyma, this family-run eatery serves traditional dishes. Try the spinach balls with Greek salad.

4 Roussiko
MAP L5 ▪ Thymiana ▪ 22710 33352 ▪ €

Located in a wonderful stone building in the small village of Thymiana, Roussiko serves classic Greek food on a rooftop terrace. The speciality is lamb or pork kotsi (a shank cut).

5 Agyra Restaurant
MAP F3 ▪ Megás Limionas ▪ 22710 32178 ▪ €

This delightful town centre ouzeri (taverna) has a quirky interior and live music playing most days. Among the dishes served here are delicious keftédes (meat balls) and shrimp pie.

6 To Apomero
MAP L5 ▪ Spiladia Kambos ▪ 22710 29675 ▪ €

Dine here for Greek dishes with a modern twist and fabulous views across the Aegean to the Turkish coast. It gets busy at weekends.

7 Hotzas
MAP L5 ▪ Kondili 3, Híos Town ▪ 22710 42787 ▪ €

Known for the wine and oúzo made by the proprietor, Hotzas has a choice of grilled and oven-cooked dishes.

Hotzas, a typical Greek taverna

8 Mesaionas Taverna
MAP K6 ▪ Mestá ▪ 22710 76050 ▪ €

Afélia (pork) and kléftiko (lamb) are just two of the delicious oven-cooked dishes on the menu at this popular taverna. Wine is local and includes souma, produced in the village.

9 Oz Cocktail Bar
MAP L5 ▪ Stoa Fragaki, Híos Town ▪ 22710 80326 ▪ €

Exquisitely plated finger-food and gourmet-style burgers are offered alongside a menu of cocktails prepared by an award-winning mixologists at this trendy eatery.

10 To Meltemaki
MAP K6 ▪ Katarráktis ▪ 22710 62105 ▪ €€

With the waves crashing almost at your feet, this beachside taverna serves fresh fish straight from the boats you can see moored at the quay.

See map on pp124–5

Sights: Lésvos

Mytilíni harbourfront

1 Mytilíni Town
MAP S2

The capital of Lésvos, Mytilíni has a multicultural feel, with fine international restaurants, waterfront bars, museums, *belle époque* churches and Venetian and Ottoman mansions.

2 Kalloní
MAP R2

At the crossroads for routes to and from the main towns, this hillside village is famous for its sardines.

3 Moní Ypsiloú
MAP Q2

Founded in the 12th century, this sprawling monastery sits on top of an extinct volcano. It has superb religious icons and an intricate wood ceiling in its *katholikon* (main church).

4 Agiásos
MAP R2

Famous as a centre for pottery (still practised today), this picturesque village is a labyrinth of tiny lanes lined with stone houses.

5 Sykaminiá
MAP R1

This village of red and white houses is perched on the slopes of Mount Lepétymnos. Strátis Myrivílis, author of the novel *The Mermaid Madonna*, which is set on Lésvos, was born here.

6 Sigrí
MAP Q2

A quiet harbour, Sigrí boasts a petrified forest – fossils of trees buried under lava for three million years.

7 Ándissa
MAP Q2

An unspoiled village, Ándissa lies near the site of a city destroyed in 168 BC by the Romans. Life centres around its café-lined square.

8 Mantamádos
MAP S1

This village of paved squares and stone houses is famous for its pottery and a rare icon of the saint Taxiarch Archangel Michael displayed in its monastery, the Moní Taxiarchón.

9 Pétra
MAP R1

A once sleepy village, Pétra is now a popular holiday spot due to its wide beach and shallow water. A volcanic monolith dominates its shore.

10 Míthymna
MAP R1

Dominated by its Byzantine castle, this village, locally called Mólyvos, is characterized by its colourful stone houses and harbour. Míthymna is believed to be the birthplace of the ancient poet Arion.

Vibrant stone houses, Míthymna

Places to Eat: Lésvos

1 Triena Café Restaurant
MAP R1 ■ Mólyvos beach
■ 22530 71351 ■ €

Enjoy breakfast, lunch or romantic evening meals at this trendy café looking out over the harbour.

2 Taverna Vafios
MAP R1 ■ Oikismós Vafiós, Vafiós ■ 22530 71752 ■ €€

Enjoy delicious Greek gourmet-style cuisine and the Lésvian dish of baked stuffed lamb with garlic at this rural taverna. Produce is from its own gardens.

The Captain's Table, beside the water

3 The Captain's Table
MAP R1 ■ Míthymna (Mólyvos) ■ 22530 71241 ■ €

Fish features prominently on the menu at this village taverna. The Captain's Table is as popular with locals as it is with visitors.

4 Kalderimi
MAP S2 ■ Thasou 2, Mytilíni Town ■ 22510 46577 ■ €

This traditional eatery on a small pedestrian street is known for its delicious local meze, making it a popular choice for lunch or dinner.

5 Cavo d'Oro
MAP Q2 ■ Harbourside, Sigrí ■ 22530 51670 ■ €

The speciality of this super fish restaurant is fish mezédes. It is located close to where the fishing boats unload their daily catch.

6 Averof
MAP S2 ■ Ermoú Street, Mytilíni Town ■ 22510 22180 ■ €

One of the oldest and most traditional tavernas in Mytilíni, the Averof serves oven-baked dishes like kléftiko (lamb), stifádo (beef) and afélia (pork). It is inexpensive and ideal for lunch.

7 Aphrodite Hotel Restaurant
MAP S2 ■ Molyvos, Mythimna ■ 22530 71725 ■ €

Open to non-residents, this hotel eatery is known for its "all-you-can-eat" buffet of traditional Greek and Lésvian dishes. Music is played most nights, while drinks comprise local wines and cocktails.

8 Tropicana
MAP R1 ■ Plateía Andreas Kyriákou, Mólyvos ■ 22530 71869 ■ €

This family-run restaurant, in a lane off a town centre square, has tables shaded under trees where you can enjoy wholesome Greek food. The signature dish is kléftiko (lamb).

9 Thalassa
MAP R1 ■ Promenade, Petra, Lésvos ■ 22530 41366 ■ €

Located in a prime position on Petra's main promenade, Thalassa specializes in seafood dishes. The shrimp saganáki is an unusual twist on a Greek favourite.

10 Orizontas Café Restaurant
MAP R1 ■ Mólyvos beach, Míthymna (Mólyvos) ■ 22530 71861 ■ €

This restaurant, with colourful linens, a cobbled terrace and palm trees, is right by the beach. It serves snacks, traditional Greek meals and drinks.

See map on pp124–5 ←

TOP 10 The Sporádes and Évvia

Statue of Aléxandros Papadiamántis, born in Skiáthos

The Sporádes, Greek for "scattered", dot the Aegean Sea in an irregular fashion. Although some isles were inhabited in antiquity, they only really came to the fore in modern times as a haunt of the rich and famous. There are 24 islands, of which four are inhabited. These are Skiáthos, which became legendary in the 1960s for the stories of celebrities seen partying on expensive yachts, the holiday island of Skópelos and the quieter Skýros and Alónissos. All four islands are known for their beaches and vineyards.

1 Skýros Town

MAP P2 ■ Archaeological Museum: Platía Brooke; 22220 91327; open 8am–3pm Tue–Sun; adm

Crowned by a large part-Byzantine, part-Venetian fortress, known as the Castle of Lykomedes, which stands on the site of an ancient acropolis,

Skýros Town is one of the prettiest capitals in the Sporádes. The town's arched lanes, lined with whitewashed cube houses, bear a striking similarity to those seen in Cycladic villages. Around the town are chapels and churches, some with blue painted domes, and interesting museums.

THE SPORÁDES AND ÉVVIA

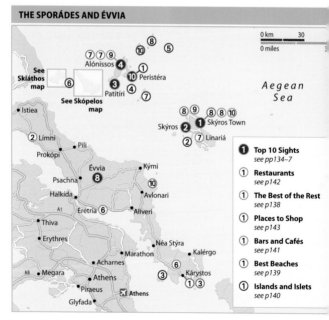

0 km 30
0 miles

Alónissos ④
⑦⑦⑨
See Skiáthos map ⑥
③ ① ⑩ Peristéra
Patitíri ④ ⑦

See Skópelos map

Istiea •

Aegean Sea

⑧⑨ ① ⑧⑧⑩
Skýros ② ① Skýros Town
② ⑦ Linariá

② Limni
Prokópi • Pili
Évvia ⑧
Psachna • Kými
Halkida • ⑩ Avlonari
A1 Erétria ⑥ Aliveri
• Thiva
• Erythres
Néa Stýra
• Marathon ⑥ Kalérgo
• Acharnes ③ Kárystos
A8 • Megara ① ③
• Piraeus Athens ✈ Athens
Glyfada •

❶	**Top 10 Sights** *see pp134–7*
①	**Restaurants** *see p142*
①	**The Best of the Rest** *see p138*
①	**Places to Shop** *see p143*
①	**Bars and Cafés** *see p141*
①	**Best Beaches** *see p139*
①	**Islands and Islets** *see p140*

The white houses of Skýros Town spread out across rocky terrain

2 Skýros
MAP P2

A dramatic rock formation greets you on approaching Skýros, its slopes covered in white houses. The largest of the Sporádes, the Island has mythological links to Achilles and King Theseus. It was colonized by Athenians, conquered by Macedonians and Romans, and used as a place of exile in Byzantine times. This now peaceful haven is famous for its own pony breed, the Skýros pony, and for furniture craftsmanship.

3 Patitíri
MAP N1

Patitíri is the capital and main port of Alónissos, replacing the nearby old capital, known as Hóra. Its pictures-que lanes are lined with traditional white houses that have red roofs and courtyards. Patitíri was built after Hóra was destroyed in the 1965 earthquake. The town radiates from the harbour, where tavernas line the seafront, fishing and pleasure craft bob in the water and ferries leave from the quayside to nearby islands.

4 Alónissos
MAP N1

Alónissos is relatively untouched by tourism. Its 3,000 or so residents live in or around Patitíri and Paleá Alónissos, with a handful living in the more rugged interior. Life here revolves around farming and working the vineyards. The island is known for its marine park, which protects rare and endangered animal species, including the monk seal *(see p55)*. The craggy limestone coastline and islets are a natural habitat for the seals.

The coastline around Alónissos

Goats grazing on a hill, Skópelos

5 Skópelos
MAP M1

Lush pine forests cover the island's two peaks, Délfi and Paloúki, and valleys of citrus groves, orchards and endless vineyards yield the delicious fruits and wines the island is famous for. Skópelos has many bays and fabulous sandy beaches, including the stunning beach of Limnonári. The main communities are located in Skópelos Town, Glóssa and the small port of Agnóndas. There are also hamlets with enchanting farmhouses known as *kalývia*.

6 Skiáthos Town
MAP M1

The main port and capital of Skiáthos island, this bustling place still retains a timeless quality. Its restaurants and bars have managed to blend in with their surroundings of red-roofed stone buildings, churches, cobbled lanes and courtyards. The town's harbour, once an important shipbuilding centre, lies between two hills. At its entrance is a pine-covered peninsula. The skyline of Skiáthos Town is dominated by churches, including the island's cathedral of Trión Lerarchón. Be sure to see the old quarter of Limniá features period merchants' houses.

7 Skiáthos
MAP M1

The island's 44 km (30 mile) coast is lined, almost exclusively, with soft sandy beaches that have made it a popular holiday destination. The most famous beach is Koukounariés, where luxury yachts are moored. A good range of hotels, restaurants and bars has steadily grown to cater to visitors, especially in the south, but despite this the Skiáthos has managed to retain its rural charm. Inland, it has forests of pine, olive groves and mountains, with rural villages, churches and deserted monasteries dotting the landscape.

8 Évvia

The fascinating archaeological site of Erétria, one of the city-states of ancient Évvia, is a must for visitors to the island. The other was Chalcis, which stood on the site of the island's modern capital, Halkída. Other places to visit include the scenic village of Prokópi and the ports of Kárystos and Kými, along with the largest spa town in Greece, Loutrá Edipsoú *(see pp34–5)*

Boats moored in the old port of Skiáthos Town in Skiáthos

9 Skópelos Town
MAP M1

The capital of Skópelos and its main port, Skópelos Town lies in a deep bay with traditional mansions and more than a hundred churches built in amphitheatre style. They, and the town's cobbled lanes, ascend to a hilltop *kástro* (fortress). The view from this Venetian castle is spectacular. Skópelos Town is also the commercial hub of the island, its fishing harbour a hive of activity. It is known for having some of the best fish restaurants on the island.

Clifftop church, Skópelos Town

10 Peristéra
MAP N1

Peristéra is one of several smaller isles and islets that lie off Alónissos, and like Skantzoúra, Pipéri and Gioúra, which are home to rare wildlife, it is now uninhabited by humans. A rugged place separated from Alónissos by two narrow stretches of water, it is popular with sailing enthusiasts who anchor off its shores. It has sandy beaches, plus the remains of an old castle to explore.

ALÉXANDROS PAPADIAMÁNTIS

One of Greece's most famous writers, Aléxandros Papadiamántis (1851–1911) was born on Skiáthos. His works, written in the country's then official language of Katharevousa, feature Skiáthos as their backdrop. They include the acclaimed *The Murderess* and the serialized *Merchants of the Nations*. He died a recluse on the island.

DAY TRIP TO ALÓNISSOS

▶ **MORNING**

A great day out from **Skiáthos** *(see p135)* or **Skópelos** is to take a ferry or hydrofoil to nearby **Alónissos** *(see p135)*. Both the scenery and the pace of life are quite different on this island. Departure times vary according to the day and season, so check in advance the destination boards on the quayside marked to **Patitíri** *(see p135)*, Alónissos. Most boats depart from 9:30 to 10:30am, so head for the quayside after a good breakfast. Boats leave from **Skiáthos Town** *(see p134)* and take from 50 minutes to more than 2 hours to reach Patitíri, depending on the kind of boat. From **Skópelos Town** the journey takes around 25 minutes.

Enjoy the coastal scenery before disembarking at **Patitíri**. Check the time of your return ferry or hydrofoil as timetables change. At Patitíri enjoy an early lunch before exploring the island. You can take a bus (timetables at the stop opposite the quay), or hire a taxi, car or bike. There are beach-hopping water taxis available too.

AFTERNOON

After lunch explore the port town of Patitíri before heading west to the old capital, **Paleá Alónissos**, which is perched precariously on a clifftop. Head north to the seaside village of **Stení Vála** *(see p138)*, where you can enjoy a bite to eat at one of its excellent tavernas. From here a road snakes towards the remote village of **Yérakas** *(see p138)*. After exploring, retrace your steps back to Patitíri for your return journey.

See map on pp134–5 ←

The Best of the Rest

1 Koliós Beach, Skiáthos
MAP M1 ■ South coast

Popular with watersports enthusiasts, Koliós is a long sandy beach that sweeps around a semicircular bay. Sailing and paragliding are on offer.

2 Troúlos Beach, Skiáthos
MAP M1 ■ South coast

This sandy beach has a gentle slope into the sea, making it ideal for families, while more secluded coves can be found around the headlands. It has a choice of tavernas and bars.

Sunbeds on the sand, Troúlos beach

3 Limnonári Beach, Skópelos
MAP M2 ■ South coast

Known for its white beach of tiny pebbles and its clear blue water, pretty Limnonári is surrounded by pine forest and has its own taverna.

4 Koukounariés Beach, Skiáthos
MAP M1 ■ South coast

With its pine forest backdrop and fine white sands, this beach is one of the most beautiful natural bays in Greece. It has restaurants and hotels nearby.

5 Megáli Ámmos Beach, Skiáthos
MAP M1 ■ East coast, Skiáthos Town

This fine sandy beach stretches around a large bay and lies adjacent to hotels, restaurants and bars. Amenities include sun beds, umbrellas and watersports.

6 Stáfylos Beach, Skópelos
MAP M2 ■ Southeast coast

This beach *(see p49)* is named after the ancient Cretan king who, according to Greek mythology, landed here to found a colony on Skópelos. The picturesque Stáfylos village is nearby.

7 Stení Vála, Alónissos
MAP N1 ■ East coast

The small coastal village of Stení Vála is a popular holiday spot, with traditional tavernas and a few hotels. Nevertheless, its beaches, including picturesque Glýfa, remain uncrowded.

8 Pouriá, Skýros
MAP P2 ■ East coast

Pouriá is a traditional coastal village known for unusual rock formations that lie just off its shoreline. A haven for snorkellers, fish can be seen swimming through sea-worn arches.

9 Yérakas, Alónissos
MAP N1 ■ North coast

This small village lies on the largely undeveloped northernmost tip of the island. Its natural beaches are best known as a refuge for the monk seal.

10 Mólos Resort, Skýros
MAP P2 ■ East coast

Scenic Mólos and its neighbour Magaziá are two of the holiday resorts on Skýros. Their beaches, including sandy Mólos and more tranquil and natural Kareflóu, are the closest to Skýros Town.

Aerial view of Mólos Resort

Best Beaches

Looking down on the sheltered Mandráki beach

1 Mandráki Beach, Skiáthos

MAP M1 ▪ West coast, near Mandráki

This very isolated beach lies in a series of secluded coves known as Eliá, Mandráki and Angistrós bays.

2 Banana Beach, Skiáthos

MAP M1 ▪ South coast, near Koukounariés

Sweeping around a large bay, this beach has amenities that include tavernas and watersports. A nudist beach, Little Banana, lies behind the rocks at one end *(see p48)*.

3 Lalária Beach, Skiáthos

MAP M1 ▪ North coast

Known for its unusual rock formations, sheer white cliffs and fine white sand, this beautiful beach can be reached only by boat.

4 Adrína Beach, Skópelos

MAP M1 ▪ Adrína, Pánormos

Head along the path from Pánormos village or by boat to get to this quiet shingle beach, which is enclosed by a pine forest *(see p51)*.

5 Glystéri Beach, Skópelos

MAP M1 ▪ North coast near Skópelos Town

This fine shingle beach lies in a small cove off a bay that protects it from strong winds. The water is excellent for swimming and there are a few tavernas here.

6 Velanió Beach, Skópelos

MAP M2 ▪ Southeast coast, near Stáfylos

A beach of golden sand lapped by clear waters, Velanió is reached by a signposted path. An official nudist beach, it is undeveloped and secluded.

7 Yérakas Beach, Alónissos

MAP N1 ▪ North coast

This often desolate and unspoiled beach is home to a research centre of the Hellenic Society for the Study and Protection of the Monk Seal.

8 Ágios Pétros Beach, Skýros

MAP P2 ▪ Atsítsa, north coast

A pretty, rocky beach that is surrounded by cedar trees, Ágios Pétros lies in a small bay. It is a popular sailing spot with locals.

9 Palamári Beach, Skýros

MAP P2 ▪ North coast

This beach is located next to the remains of the ancient town of Palamári. Quiet and peaceful, it has golden sand, clear water and a few sleepy tavernas nearby.

10 Ochthoniá Beach, Évvia

MAP N3 ▪ Northeast coast

Usually deserted, this wild and exposed beach lies near the rocky cliffs, rich vegetation and windswept trees of Cape Ochthoniá.

See map on pp134–5

Islands and Islets

1 Peristéra Island

This small island *(see p137)* lies off Alónissos and it is part of the Sporádes Marine Park. Its sandy coves attract the endangered monk seal *(see p55)*.

2 Valáxa Island
MAP P2 ■ Off Skýros

This uninhabited island and its neighbouring Skyrópoulo and Erínia islets attract yachtsmen because of their sheltered coves. They are accessible by boat from Péfkos.

3 Petáli Island
MAP M2 ■ Off Évvia

The rugged island of Petáli is said to have been owned by royalty and several celebrities, including the Spanish painter Pablo Picasso, whose house still stands in the forest near the beach.

4 Dýo Adélfia
MAP N1 ■ Near Alónissos

The twin islets of Mikró and Megálo Adélfia are known for their great scuba-diving spots. However, the sport is strictly controlled as the islets lie in the Sporádes Marine Park.

5 Pipéri Island
MAP P1 ■ Near Alónissos

Covered by pine forest, this island lies in the Sporádes Marine Park. It is home to several rare species of flora and fauna, including the endangered Eleonora's falcon *(see p55)*. Disembarking here is pro-hibited, but visitors can view the amazing wildlife from a boat.

6 Árgos Island
MAP M1 ■ Off Skiáthos

Árgos can only be reached by private boat and is a popular anchorage for yachts sailing around the Sporádes. The views of Skiáthos Town from the island's shores are outstanding.

7 Skantzoúra
MAP N2 ■ Near Alónissos

This small archipelago of six islets is most famous for its beautiful cedars and its sacred monastery, the Moní Evangelístria, which lies on its main island. Hellenistic remains suggest ancient occupation.

8 Gioúra Island
MAP N1 ■ Off Alónissos

Uninhabited, the island of Gioúra is part of the Sporádes Marine Park. This small, rugged piece of land has archaeological remains that suggest it was once a settlement inhabited as far back as Neolithic times.

9 Tsougriá Island
MAP M2 ■ Off Skiáthos

Along with its islet, Tsougriáki, this richly forested island has a couple of sandy beaches within protected bays and is a popular day-trip destination for visitors staying in Skiáthos Town.

10 Kyrá Panagía Island
MAP N1 ■ Off Alónissos

This tiny island has the remains of a settlement, which is believed to be ancient Alónissos. Today there are just a handful of residents, who have built a community that is administered from Alónissos.

Boats anchored, Kyrá Panagía Island

Bars and Cafés

1 **Espresso Bar, Skópelos**
MAP M1 ▪ Pánormos Beach
▪ 24240 23951

Set right on the beach, the Espresso Bar is about relaxing over a coffee or tucking into a light meal. It's a great place to watch the sunset over an evening cocktail.

2 **Pillows Cocktail Bar, Skiáthos**
MAP M1 ▪ Old Port, Skiáthos
▪ 06987 411711

This bar serves generous cocktails at reasonable prices. The owner, Panos, lays on music and dancing in the evenings.

Seating at Pillows Cocktail Bar

3 **Panorama Pizza, Skiáthos**
MAP M1 ▪ Ag Anargyri ▪ 24270 22877
A great hilltop pizzeria where delicious pasta, handmade pizzas with varied toppings, garlic bread and speciality coffees can be enjoyed while looking out over Skiáthos Town.

4 **Mercurius Bar, Skópelos**
MAP M1 ▪ Harbourside, Skópelos Town ▪ 24240 24593
With a lovely terrace overlooking Skópelos harbour, this trendy bar is known for its creative cocktails and music and cultural events.

5 **Rock and Roll Bar, Skiáthos**
MAP M1 ▪ Old Port, Skiáthos Town
▪ 24270 22944

This bar plays rock, dance and Latin music and has cushioned seating in a courtyard overlooking the harbour. Drinks include cocktails, beer and local spirits.

6 **Enzo Cafe-Bar, Eretria**
MAP D2 ▪ Arxaiou Theatrou 1, Erétria ▪ 69487 15389
Fresh juices, coffee and snacks are served here during the day. After dark, the atmosphere gets lively and there are special events throughout the summer.

7 **Kavos Lounge Bar, Skyros**
MAP P2 ▪ Linaria harbour
▪ 22220 93213
Set right on the harbourside of Linaria village, Kavos couldn't offer a better spot to linger. Enjoy local wines, cocktails and snacks on its terrace.

8 **Rock n Roll Bar, Skiáthos**
MAP M1 ▪ Old port, Skiáthos Town ▪ 24270 22944
This happening bar overlooks the sea, with bobbing traditional fishing boats making for a picturesque view. The DJ plays a range of music genres, including rock and dance tracks.

9 **Oionos Blue Bar, Skópelos**
MAP M1 ▪ Off Platía Platanos, Skópelos Town ▪ 69424 06136
An atmospheric lounge bar with a heady mix of ethnic music, jazz, blues and soul. Light meals, beer, cocktails, wine and whisky are served here.

10 **Platanos Jazz Bar, Skópelos**
MAP M1 ▪ Platía Plátanos, Skópelos Town ▪ 24240 23661
Dark and elegant decor, sophisticated cocktails and live music from jazz to Latin American and blues, are what make Platanos a popular nightspot.

See map on pp134–5

Restaurants

(1) Garden Restaurant Alexander, Skópelos

MAP M1 ▪ Skópelos Town ▪ 24240
22324 ▪ €€

Set in the main square, this restaurant
is known for its garden. Enjoy a meal
of classic local dishes and wine,
while listening to Greek music.

(2) Astron, Évvia

MAP D2 ▪ Limni ▪ 22270 31487
▪ €€

This popular restaurant serves local
meat, seafood and vegetarian dishes
and homemade desserts. It offers
wines from around the world.

(3) Infinity Blue Restaurant, Skiáthos

MAP M1 ▪ Harbourside, Koliós
▪ 24270 49750 ▪ €€

Swordfish, sea bream and lobster
are some of the seafood dishes
offered at this elegant restaurant.
Classic Greek starters, mains,
desserts and more than 50 wines
add to the experience.

(4) Restaurant Terpsis, Skópelos

MAP M1 ▪ Stáfylos beach ▪ 24240
22053 ▪ €€

Meaning "pleasure" in Greek, Terpsis
was founded in 1965 and is famous
for its spicy chicken stuffed with
walnuts from the trees in its garden.
Greek music plays as you dine.

(5) Korali Restaurant, Skópelos

MAP M1 ▪ Agnóndas ▪ 24240
22407 ▪ €

This beachside taverna has a
magical setting and its menu is
classic Greek. Mains include *afélia*
(pork) and *kléftiko* (lamb).

(6) To Choni, Évvia

MAP D3 ▪ Kárystos ▪ 22240
24152 ▪ €€

Set in a stone-built cottage, this
country restaurant serves traditional
Greek dishes – most are homemade,
including the *tyropitára* (cheese pies).

(7) Agnanti Restaurant, Skópelos

MAP M1 ▪ Glóssa ▪ 24240 33606 ▪ €€

Founded in 1953, Agnanti serves
classic Greek dishes with a modern
twist, using fresh and organic produce.

(8) Anemos Restaurant, Skiáthos

MAP M1 ▪ Harbourside, Skiáthos
Town ▪ 24270 21003 ▪ €€

The oldest taverna in town, Anemos
offers a Greek menu with fine local
wines. The harbour view is spectacular from here, especially at night.

(9) The Muses, Skópelos

MAP M1 ▪ Seafront, Skópelos
Town ▪ 24240 24414 ▪ €€

Dine inside this pretty seafront
restaurant, where artworks and
tropical plants feature, or outside
on its esplanade. Fish and meat
dishes are straight from the grill.

(10) Windmill Restaurant, Skiáthos

MAP M1 ▪ Harbourside, Skiáthos
Town ▪ 24270 24550 ▪ €€

Housed in a restored 19th-century
windmill, this romantic restaurant
serves fine local dishes on its terraces overlooking the harbour.

View from the Windmill Restaurant

Places to Shop

Baklava, for sale at Roukliotis Bakery

1 Roukliotis Bakery, Évvia
MAP E3 ▪ Kárystos ▪ 22240 23290

Roukliotis bakes its own pastries, bread and confectionery. Try the *bougatsa*, a traditional greek dessert of custard wrapped in filo pastry and garnished with sugar and cinnamon.

2 Rodios, Skópelos
MAP M1 ▪ Skópelos Town ▪ 24240 22924

This shop sells everything from local artifacts to works of art, ceramics and local handicrafts. Look out for the woven rugs and embroidered linen.

3 Makri Kondilia, Évvia
MAP E3 ▪ Saxtouri 77, Kárystos ▪ 22240 26300

This beauty centre stocks a range of Geothermica spa products made locally using raw materials from the natural thermal springs of Loutrá Aidipsoú in northern Évvia.

4 The Blue House Art Gallery, Skiáthos
MAP M1 ▪ Old Port, Skiáthos Town ▪ 24270 21681

Housed in a traditional Skiáthian building with blue painted shutters, this delightful gallery has displays of individually designed jewellery.

5 Kactri, Skópelos
MAP M1 ▪ Skópelos Town ▪ 24240 24592

Come here for unique handcrafted jewellery made with gold, sterling silver and precious gemstones.

6 Driftwood Art, Skiáthos
MAP M1 ▪ Panora, Skiáthos Town ▪ 24270 22296

In a delightful building on a cobbled lane, this shop has some remarkable *objets d'art* made from recycled wood.

7 Rodios Pottery, Skópelos
MAP M1 ▪ Port ▪ 24240 23605

The Rodios family has been creating pottery since the 1920s using a technique that dates back to ancient potters of 4th and 5th centuries BC.

8 Faltaits Museum's Shop
MAP P2 ▪ Palaiopyrgos, Skýros ▪ 22220 91232

Visitors can choose from a variety of souvenirs, such as handmade ceramics, Skyrian embroidery, copies of Manos Faltaits' artworks, books and other memorabilia at this shop.

Displays of jewellery, Galerie Varsakis

9 Galerie Varsakis, Skiáthos
MAP M1 ▪ Papadiamánti Street, Skiáthos Town ▪ 24270 22255

This old art and antiques shop has icons, embroidery, textiles and jewellery. The basement resembles a traditional kitchen, with antique pots.

10 Símos Jewellery, Skiáthos
MAP M1 ▪ Papadiamánti Street, Skiáthos Town ▪ 24270 23777

Símos specializes in 14- and 18-carat yellow and white gold jewellery in modern or traditional designs. Many of the pieces are made by the owner.

See map on pp134–5

🔟 The Argo-Saronic Islands

Comprising the main islands of Égina, Hydra, Salamína, Póros and Spétses, with the traditionally Ionian Kýthira further south, the Argo-Saronics stretch some 200 km (125 miles) from Piraeus in Athens to the Peloponnese coast. Smaller islets, such as Angístri, Moní, Trikéri and Spetsopoúla beg to be explored. The archipelago has a refined feel in many of its towns, and yet by contrast, small sailing caïques and mules form the only means of transport on some islands.

Statue, Battle of Salamis

THE ARGO-SARONIC ISLANDS

Moní Faneroménis ⑦
Corinth
Salamína ⑧ ⑧ Ambelákia
Athens
⑤ ⑨

Póros

0 km 3
0 miles 3

③ ⑤ ⑦ ③
Angístri ⑤
See Égina map
⑥ Temple of Poseidon
⑦
Neórion ⑥
Áskeli
Argos
Nafplio
⑩
Galatás
Póros Town
④ ⑥ ⑩ ④ ⑧ ⑩
⑨

See Póros map

Ermióni
⑥ ④
⑩ Hydra Town
Porto Heli ⑨
⑨ ① ⑩ ⑤
See Spétses map
② Hydra ② ⑤ ⑧ ⑨
⑥ ⑤
⑥ ⑤

Spétses

⑦
Ligoneri ① ③
Kounoupitsa
Ágii ⑧
Anárgyroi
Spétses Town
③ ⑥ ⑦ ⑦
⑨
① Agía Marí

Leonídi

0 km 2
0 miles 2

Geráki

Aegean Sea

Molai

Monemvasía

Neápoli
①

Agia Pelagia
☒ ⑦
Kýthira ③ ⑩
⑩
④
↘ ⑧ 30km

①	Top 10 Sights *see pp145–7*	①	Bars, Cafés and Clubs *see p151*
①	Restaurants *see p152*	①	Best Beaches *see p149*
①	Islands and Islets *see p150*	①	The Best of the Rest *see p148*
①	Places to Shop *see p153*		

0 km 25
0 miles 25

The Panagitsa church, Égina Town

1 Égina
MAP L1

Famous as the place where the world's first ever coins were minted in 700 BC and home to the impressive Temple of Aphaia *(see pp32–3)*, Égina is a pine-covered island with picturesque bays and beaches. It's captial, Égina Town, hugs a natural harbour and multiple ferries from Athens arrive here daily. The island has enjoyed a lively past and much prosperity as a maritime trading centre. Today, life revolves around the production of pistachio nuts, considered the best in Greece.

2 Spétses
MAP D4

Spétses is characterized by the scent of the pine forest that was planted in the early 1900s. The interior is dotted with holiday homes. Spétses Town, its capital, is the largest community and spreads along the eastern coastline for 2 km (1 mile). Here, life revolves around Dápia quay, where fashionable tavernas look out over the bay. The Ágios Nikólaos church and 19th-century mansions give a hint of the town's past wealth, largely made from shipbuilding.

3 Kýthira
MAP C5

Arriving on Kýthira feels like stepping back in time. Tiny lanes flanked by dry-stone walls link hamlets with blue and white homes, many show-ing elements of Venetian architec-ture. South of the Peloponnese, Kýthira was once governed by the Venetians and British as part of the Ionians, but is now an Argo-Saronic island. Many residents are returning Australian Kythirans, who have given the island a distinctive character.

Égina **1**

| n | 3 |
| iles | 3 |

Souvála ②   Vaia

Kypséli ④   ⑦   Mesagrós ⑧ ③

⑤ Kontos   Agía Marína ①②⑧

See Égina Town map below   ①⑩ ②⑨

③   ⑩ Pahiá Ráhi

③⑥⑥   ④   Anítseo

Pérdika   ①   Sfikári

Égina Town ①

DIMITRÍOU PETRÉS   KAPPOU

⑦ ⑨

THOMAIDOU

③
⑨ ④ ④ ④
⑥② ④
②
⑧
Égina Harbour   SPYROU RODI

⑤
tres 200
ds 200   ⑤ ← 150 metres

① **Restaurants: Égina** *see p155*

① **Sights: Égina** *see p154*

Kýthira's rugged landscape

Póros Town on the island of Sfairiá, Póros

4 Póros
MAP K3

Póros is made up of two islands: the richly forested but almost uninhabited Kalávria to the north and Sfairiá to the south. They are connected by a causeway. Its harbourside capital, Póros Town, lies on Sfairiá and despite the rapid growth of its tourism industry has managed to retain its traditional charm. Tiny streets, lined with pastel-coloured 19th-century houses clinging to the hillside, give the town its character.

5 Angístri
MAP J2

With a timeless charm and verdant landscape of pine trees, Angístri is a popular summer haunt of Athenian weekenders. Its main port is Skála, where there are a handful of hotels and restaurants, and the island's best beach, which runs along the coast from its church. In the past, the island enjoyed reflected glory and wealth from Égina, and excavations at the sites of Megarítissa, Apónissos and ancient Metochí have revealed many valuable treasures.

6 Temple of Poseidon, Póros
MAP K3 ■ Kalávria

Set near the coast, in the heart of Kalávria, are the ruins of the Temple of Poseidon. Believed to have been built around 520 BC, the temple was a refuge for Kalavrian tribes. Built in Doric style, it would originally have had 12 columns along its longer sides and 6 along its shorter sides. The site is said to have been where in 322 BC the orator Demosthenes poisoned himself rather than be captured by the Macedonians.

7 Moní Faneroménis, Salamína
MAP D3 ■ Salamína ■ 21046 81861 ■ Open 7am–12:30pm and 3:30pm–sunset daily ■ Donation

The 17th-century Moní Faneroménis is on the north coast of Salamína, on the spot where an 18th-century icon, known as the Noefaneísis Panagía, was discovered by St Lauréntos. It depicts the *Last Judgement*. The monastery was used during the War of Independence (1821–9) as a hospital and refuge. In 1944, it was converted into a convent. The nuns host a festival every year on 23 August in honour of the Virgin Mary.

AJAX

A mythological Greek hero, Ajax was made famous in Homer's poems *Iliad* and *Odyssey*. He is said to have been born on the island of Salamína. The King of Salamis, legend has it, Ajax was a man of great strength and a fearsome warrior.

8 Salamína
MAP D3

The largest of the Saronic Gulf islands and separated from mainland Greece by a 2-km- (1-mile-) wide strait, Salamína is famous for being the site of the historic Battle of Salamis in 480 BC. The battle between the victorious Greek fleet and the Persian Empire was fought in the strait, with the Salaminians playing a key role. The island's capital, Salamína Town, lies in a deep bay to the west, while Ágios Nikólaos is a popular holiday spot.

9 Hydra
MAP D4

Hydra rose to fame in the 1960s when two films were shot here: *Boy on a Dolphin* starring Sophia Loren and the comedy *Island of Love*. Pretty whitewashed mansions belonging to artists, writers and the Athenian elite cling to the hillside rising steeply from the harbour. Road-free, it is regarded as an exclusive holiday destination.

Boats fill the harbour, Hydra Town

10 Hydra Town
MAP D4 ■ Ýdra

Hydra Town, capital of Hydra, is a mix of harbourside tavernas, frequented by a chic clientele, and elegant pastel-coloured homes. Built by mariners who had amassed wealth during the Napoleonic Wars by building block-ades, these 19th-century mansions, known as *archontiká*, are built of local stone. One mansion houses a marine academy and one a school of fine arts.

DAY TRIP FROM ÉGINA TO ANGÍSTRI

▶ MORNING

Many ferries run between the northern Argo-Saronic islands daily, with one of the most popular day trips being between Égina (see p145) and Angístri. Start your day with a good break-fast and head to the harbourside at Égina Town (see p154). The harbour is easy to find and big boards detail the ferry departure times. Look out for the one marked Skála, Angístri Island. The first ferry leaves at about 9:30am and then every couple of hours, but times do vary so be sure to check. Caïques also ply the waters at irregular times. The ferry crossing takes about 20 minutes. Look out for **Moní island** (see p150) to your left and **Metopí islet** (see p150) to your right, and then look ahead to catch your first glimpse of picturesque **Skála**. Once you have disem-barked, stop for coffee before making your way to Skála beach (see p149). You can also hire a scooter or bike to explore the island. If you are lucky, you may be able to catch one of the infre-quent buses that run from Skála to **Limenária** via the small port of **Megalochóri** (see p148). All the villages en route have tavernas for lunch.

AFTERNOON

On the steep hillside above Skála, **Metochí** provides a sense of the island's history and breathtaking views. Retrace your steps back to Skála in time to catch a return ferry. They leave around 5:40pm, but check as times can vary. Back in Égina you can enjoy a delicious evening meal at one of the har-bourside restaurants (see p155).

See map on pp144–5 ←

The Best of the Rest

1 Sárpa Bay, Égina
MAP K1 ▪ Égina Town

With its crystal-clear waters, this bay off the road to Pérdika is one of the capital's best beaches. It has a café and golden sand dotted with tropical-style umbrellas.

2 Souvála, Égina
MAP L1 ▪ Near Vathý

Famous for its mineral-rich springs, the once sleepy fishing village of Souvála is now Égina's second port. Ferries arrive here from Athens. It is a picturesque spot with tavernas lining its esplanade.

3 Megalochóri, Angístri
MAP J2

The capital of the island of Angístri, Megalochóri (Mýlos) is an unspoiled coastal village of blue and white stone houses. These houses cluster along tiny streets that give no hint of this village's former importance as a key naval base.

4 Vlychós, Hydra
MAP D4

A tiny village that revolves around its enchanting blue and white church, Vlychós has a pretty beach dotted with tropical-style sun umbrellas. A handful of tavernas and a small resort are close by.

5 Kamíni, Hydra
MAP D4

Known as Mikró Kamíni, for its diminutive size, this picturesque fishing harbour has its own castle, a collection of traditional tavernas and a quiet beach.

6 Bísti Bay, Hydra
MAP D4

Bísti bay, with its pebbly beach and safe water, is a lovely spot, especially for watersports. Kayaking, snorkelling and scuba diving are available.

7 Zogeriá Bay, Spétses
MAP D4

This bay, the most beautiful on Spétses, is the island's best-kept secret. It lies in a small cove surrounded by a dense, scented pine forest. Its beach is sandy and the water is sapphire blue.

8 Ambelákia, Salamína
MAP D3

This is the oldest village on the island, and is famous as the site of the Battle of Salamis (see p147). It boasts a citadel, a port and the ancient ruins of an acropolis.

9 Boúrtzi, Póros
MAP K3

A tiny fortified island off Póros, Boúrtzi has a chequered past. Under the rule of Basil I the Macedon (AD 812–886), the island was the scene of battles against the Saracens and was later used by the Germans as a refuge during World War II.

10 Venetian Castles, Kýthira

Kýthira has several fortifications, including a mighty Venetian castle that stands on rocks overlooking its capital. Nearby are the ruins of the Kato Hora castle with its well-preserved Lion of St Mark statue and the Avlémonas bastion.

Kamíni, with its seashore castle

Best Beaches

Agía Marína beach, Spétses

1 Agía Marína, Spétses
MAP D4 ▪ Near Dápia

Famous as much for its beauty as for the remains of an Early Bronze Age civilization, this beach has watersports and sun beds *(see p50)*. Inland is a chapel of the same name.

2 Agía Marína Beach, Égina

The fabulous beach at this small resort is popular with wind surfers and sailing enthusiasts. The village itself is picturesque with its white-washed houses and authentic tavernas *(see pp154 and 155)*.

3 Marathónas Beach, Égina

MAP K1 ▪ South of Égina Town

Busy yet scenic, the beach alongside the fishing village of Marathónas is known for its excellent selection of fish restaurants. With sun umbrellas, snorkelling and diving available, it is a popular spot.

4 Aiginítissa Beach, Égina

MAP K1 ▪ South of Égina Town

Much quieter than its neighbour, Marathónas, this beach has shallow, crystal-clear water ideal for children. Loungers and parasols are available.

5 Skála Beach, Angístri
MAP K2

The longest stretch of beach on the island and its sandiest, Skála attracts both locals and day-trippers from Égina. It runs along the coast from the town's Ágii Anárgyroi church.

6 Neórion Beach, Póros
MAP K1 ▪ Northwest of Póros Town

Sandy and surrounded by tall pine trees, this is one of the most picturesque beaches on the island. Sometimes known as Mikró Neório, it has a good selection of tavernas, as well as watersports to enjoy.

7 Russian Bay, Póros
MAP K1 ▪ Northwest of Póros Town

Deriving its name from when it harboured Russian ships that aided the Greek fleet during the revolution of the 1820s, this sheltered bay has golden sand and a few tavernas.

8 Agía Paraskeví Beach, Spétses

MAP D4 ▪ Near Dápia

The beautiful church of Agía Paraskeví lends its name to this long stretch of sand and pebble enclosed within a cove. Pine trees provide it with shelter.

9 Ágii Anárgyroi Beach, Spétses

MAP D4 ▪ Near Dápia

Of the many good beaches in the Argo-Saronics, the Ágii Anárgyroi is the largest. Lying in a cove, this popular sandy beach has shallow waters for safe swimming.

10 Palaiópoli Beach, Kýthira
MAP D5 ▪ East coast

Kýthira's largest beach and, legend has it, where the goddess Aphrodite was born, Palaiópoli has tavernas, sun umbrellas and watersports. It is on the site of the ancient city of Scandeia.

See map on pp144–5

Islands and Islets

Sand dunes, Elafónissos Island

① Elafónissos Island
MAP C5 ■ North of Kýthira

Visitors from the Peloponnese make a short sea crossing to reach this almost barren island with light-coloured sand dunes stretching as far as the eye can see. Its main settlement is the small port.

② Trikéri and Spetsopoúla Islands
MAP D4 ■ East of Spétses

Lying off the coast of Spétses, the tiny, somewhat barren island of Trikéri and the privately owned Spetsopoúla island are popular landmarks with sailors. Although there are no public moorings, yachts often sail in their waters.

③ Metopí Islet
MAP K2 ■ Between Égina and Angístri

This little islet, dominated by its church, offers tranquillity and just nature for company. It has some fine beaches and moorings, making it a popular day-trip from Égina or Skála.

④ Nafítilos Islet
MAP D4 ■ South of Kýthira

Nafítilos, along with a cluster of other tiny islets, including Psíra, Poretí, Thérmones and Póri, lie off Kýthira's coast. All are uninhabited and are popular with day-trippers due to their glorious beaches.

⑤ Alexandros Island
MAP D4 ■ Southwest of Hydra

Often known as Platoníssi Nisída, Aléxandros is one of the many islands lying south of Hydra. Undeveloped, it has pretty coves and beaches.

⑥ Petássi Island
MAP D4 ■ Northwest of Hydra

Popular with yachtsmen, this small island offers safe anchorages in the sheltered Bísti bay. There are many coves along its indented coastline.

⑦ Makroníssi Island
MAP D4 ■ Off Diakófti

Reached by a causeway from Diakófti, this flat island has rich vegetation, lots of birdlife, sandy beaches and an extremely picturesque port.

Fishing boat moored at Makroníssi

⑧ Andikýthira Island
MAP D5 ■ South of Kýthira

This small, isolated island has a population of around 50, all of whom live in the three villages of Potamós, Galaniána and Harhaliána.

⑨ Dokós Island
MAP D4 ■ North of Hydra

Inhabited since Neolithic times and with valuable Hellenic and Byzantine archaeological remains, this remote island is home to Orthodox monks.

⑩ Moní Island
MAP K2 ■ Between Égina and Angístri

Reached from Pérdika by boat, this pine-covered island – inhabited only by deer and goats – is popular with day-trippers for its beaches.

Bars, Cafés and Clubs

1 **Thymari, Égina**
MAP L1 ■ Afeas, Agía Marína
■ 22970 32859

This trendy restaurant, open all day from breakfast through to dinner, is famed for its *moussakás* and pizzas, plus an extensive selection of Greek starters that are a meal on their own.

2 **Barracuda Beach Bar, Égina**
MAP L1 ■ Agía Marína ■ 22970 32095

Serving colourful cocktails, snacks and American-style waffles, this beach bar is a great place for a bite.

3 **Bluefield Burger, Spétses**
MAP D4 ■ Plateía Rologiou, Spétses Town ■ 22980 73388

This lively place with courtyard dining is a magnet for locals as well as visitors, and is known for its range of gourmet-style burgers. Each burger is a work of art, accompanied by fries and imaginative salads.

4 **En Plo, Égina**
MAP K1 ■ Harbourside, Égina Town ■ 22970 26482

Serving waffles and crêpes with a range of beverages, this lively café is right by the harbour. It is open for breakfast, lunch and informal dinners.

5 **Remvi, Égina**
MAP K1 ■ Near the harbour, Égina Town ■ 22970 28605

Open all day and night, and located right on the harbour, this is the place to enjoy coffee, drinks and snacks during the day and cocktails and music until the early hours.

6 **Mourayo Music Bar, Spétses**
MAP K1 ■ Old Harbour, Spétses Town ■ 22980 73700

Housed in a 19th-century building, Mourayo has been at the heart of Spétses nightlife since 1975. Drinks and music can be enjoyed on the large open-air terrace.

7 **Club Stavento, Spétses**
MAP D4 ■ Old Harbour, Spétses Town ■ 69775 04171

With its indie, garage and hip hop sounds, this lively nightclub attracts the young and trendy. DJs and live bands play into the night.

8 **Pita Tom, Égina**
MAP L1 ■ Agía Marína
■ 22970 32787

This central fast-food eatery is famous for its pork and chicken *souvlaki*, which can be cubed and skewered, or cooked donar kebab-style *(gyros)* and served in freshly-baked pitta bread with fries and salad.

9 **Yacht Club Panagakis, Égina**
MAP K1 ■ Dimokratías 20, Égina Town ■ 22970 26654

This stylish seafront café-bar serves coffee with sweet and savory crêpes and grills and pizzas. Its bar has an extensive cocktail menu.

Trendy decor, Yacht Club Panagakis

10 **Dionysos Café Bar, Póros**
MAP K3 ■ Póros Town beach
■ 22980 23511

Open all day for snacks and drinks, the Dionysos is a welcoming café-bar. R&B and the latest pop tunes are played into the night.

See map on pp144–5

Restaurants

PRICE CATEGORIES

For a three-course meal for one with half
a bottle of wine (or equivalent meal),
taxes and extra charges.

€ under €30 €€ €30–€50 €€€ over €50

1 Patralis Fish Tavern, Spétses

MAP D4 ▪ Kounoupitsa seafront
▪ 22980 75380 ▪ €

Dating back to 1935, this taverna was
established by a local fisherman and
it continues to specialize in fresh fish.

2 Kodylenia Taverna, Hydra

MAP D4 ▪ Seafront, Kamínia
▪ 22980 53520 ▪ €

This charming taverna, a few steps
from the water's edge, is owned by a
fisherman. He makes the daily catch,
which is then beautifully prepared.

3 Akroyialia Restaurant, Spétses

MAP D4 ▪ Kounoupitsa seafront
▪ 22980 74749 ▪ €

Dining on the beachside terrace of
this restaurant is magical. Some of
the tables are set right on the beach.

4 Tavern Karavolos, Póros

MAP K3 ▪ Harbour, Póros Town
▪ 22980 26158 ▪ €

With nostalgic artifacts adorning its
blue and white walls, this little taverna
serves classic Greek dishes. Try its
mezédes, some of the best in Póros.

The dining room at Tavern Karavolos

5 Christina's Taverna, Hydra

MAP D4 ▪ Spilios Charamis, Hydra
Town ▪ 22980 53516 ▪ €

Grilled seafood is the speciality here. If
you feel adventurous, try the scorpion
fish. Dining is on the rooftop terrace.

6 Taverna Poseidon, Póros

MAP K3 ▪ Plateia Virvili, Póros
Town ▪ 22980 23597 ▪ €€

Fresh fish caught by the owner and
meats cooked over charcoal are
among the temptations at this
friendly spot right on the seafront.

7 Toxotis Taverna, Angístri

MAP K2 ▪ Skála crossroad,
Skála ▪ 22970 91283 ▪ €€

Serving both Greek and international
dishes, and a good selection of
vegetarian options, this charming
taverna also offers local wines.

8 Sunset Restaurant, Hydra

MAP D4 ▪ Harbour, Hydra Town
▪ 22980 52067 ▪ €€

As its name suggests, diners at this
restaurant enjoy stunning sunsets.
Its menu is traditional Greek with
a contemporary twist, and is
accompanied by a fine wine list.

9 Hydronetta, Hydra

MAP D4 ▪ Boudouri, Hydra
Town ▪ 22980 54160 ▪ €€

In an idyllic spot perched on rocks
above a bay, this taverna presents
fish and seafood with panache.
Cocktails are equally imaginative
and there are great
sunset views.

10 Taverna Rota, Póros

MAP K3 ▪ Harbour, Póros
Town ▪ 22980 25627 ▪ €

With several tables set
out under canopies,
this pleasing restaurant
serves breakfasts and
classic Greek meals.

Places to Shop

Clothes on display, Speak Out

⑤ Ergastiri Vaxevanidi, Salamína

MAP D3 ■ Faneroménis 88 ■ 21046 50365

Colourful dresses, beachwear and sandals, bags and pretty bead necklaces are some of the items sold here.

⑥ Anchor, Égina

MAP K1 ■ Harbourside, Égina Town ■ 22970 26422

Easy to find on the seafront and housed in an attractive stone building, Anchor offers the chance to buy an elegant evening outfit or more casual clothing as well as accessories.

① Speak Out, Hydra

MAP D4 ■ Harbourside, Hydra Town ■ 22980 52099

Housed in a period property along Hydra Town's seafront, this boutique is full of eclectic fashion goods that vie for attention with its *objets d'art*.

② Kendro Tipou, Égina

MAP K1 ■ Harbourside, Égina Town ■ 22970 27742

This attractive little bookshop, overlooking the harbour, has a wide range of novels and guides to help you around the island. It also sells Greek and international newspapers.

⑦ Koukiaris Konstantinos, Spétses

MAP D4 ■ Spétses Town ■ 22980 29468

Spétses is well known for its gold and silver jewellery, and this shop sells an elegant selection of pieces.

⑧ Stoa, Póros

MAP K3 ■ Póros Town beach ■ 22980 26761

Stoa sells leather bags and belts, acrylic paintings of local scenes, statuettes and pottery pieces, many crafted by local artisans.

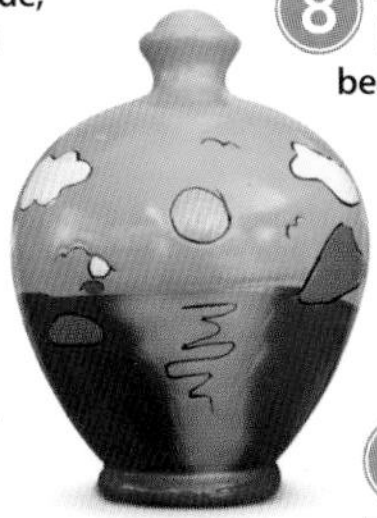

Pottery for sale at Stoa

③ Asao, Égina

MAP K2 ■ P. Irioti 21, Égina Town ■ 22970 27280

With its colourful selection of upmarket holiday wear, sandals, bags and pashminas, plus unusual items of handmade jewellery, this is a place to indulge yourself or to find the perfect gift.

⑨ Exantas, Salamína

MAP D3 ■ Papaníkoloaou Evangélou 6 ■ 21046 51886

This shop is a veritable Aladdin's cave of unusual souvenirs. Small items of handcrafted furniture sit beside metal lanterns, children's toys and colourful ceramics.

④ Gion, Póros

MAP K3 ■ Harbour, Póros Town ■ 22980 24158

Gion stocks designer beachwear, sportswear names Timberland and Nike, and holiday accessories for men, women and children.

⑩ Tsagkari Pastry Shop, Hydra

MAP D4 ■ Hydra Town ■ 22980 52314

Only a few feet from both the central market and the port, this shop occupies an enviable location. The famous roasted or boiled almond sweets of Hydra are made here.

See map on pp144–5

Sights: Égina

Fishing boats hug the harbour, enhanced by the Panagitsa church, Égina Town

1 Égina Town
MAP K1

Neo-Classical buildings in pastel shades, often occupied by fish tavernas, line this town's picturesque harbour. Fishing boats crowd its waterside, while landmarks include the ruins of the Temple of Apollo.

2 Temple of Apollo
MAP K1

Just one column is left of this great 6th-century-BC temple built in honour of Apollo, prompting it to be known as Kolóna ("the Column").

3 Temple of Aphaia
This well-preserved temple of goddess Aphaia is more than 2,500 years old (see pp32–3).

4 Kypséli
MAP K1

Pistachio plantations stretch along the Kypséli plateau, producing the island's biggest export.

5 Monastery of Ágios Nektários
MAP K1

This monastery houses one of the largest Greek Orthodox cathedrals in the world. It was built in the early 1900s by the patron saint of the island, Nektários, who is buried here.

6 Pérdika
MAP K2

The quaint fishing harbour of Pérdika is a gem. It is lined with fish tavernas where you can sit and watch fishing vessels and yachts. Surrounded by nature, the town is a walker's heaven.

7 Paleohóra
MAP L1

For around 1,000 years until the early 19th century, this deserted village was Égina's capital. It is said that its inhabitants were forced into slavery by pirates. Visitors can see the remains of various churches here.

8 Mesagrós
MAP L1

Known for its ceramics, this mountain village is a popular tourist spot. It is home to traditional 19th-century houses and a pretty church.

9 Agía Marína
MAP L1

This cosmopolitan holiday resort with a sandy beach (see p149) takes its name from the church in the bay.

10 Pahiá Ráhi
MAP L2 ▪ www.ekpazp.gr

This rural mountain village is home to the Hellenic Wildlife Hospital of Égina, a refuge for sick animals.

Restaurants: Égina

PRICE CATEGORIES

For a three-course meal for one with half a bottle of wine (or equivalent meal), taxes and extra charges.

€ under €30   €€ €30–€50   €€€ over €50

(1) Akrogiali Café
MAP L1 ▪ Agía Marína Beach ▪ 22970 32427 ▪ €€

Family-run Akrogiali opened way back in 1947 and is known for its lively atmosphere and great cuisine. Right beside the beach, its platters of squid, shrimps and octopus are legendary.

(2) Miltos
MAP K2 ▪ Harbourside, Pérdika ▪ 22970 61051 ▪ €€

Overlooking a fishing harbour, this restaurant serves classic Greek fare, such as *mezédes* of dips followed by meat and fish mains.

Dining next to the sea, Nontas

(3) Nontas
MAP L1 ▪ Pérdika ▪ 22970 61233 ▪ €

This pretty taverna has terraces next to the fishing harbour. Greek dishes and mouthwatering seafood, caught fresh that day, feature on the menu.

(4) Agora Tavern
MAP K1 ▪ Fish market, Égina Town ▪ 22970 27308 ▪ €

This chic little restaurant is popular with locals. Octopus, sea bream and shrimp are prominent on the menu.

(5) Babis
MAP K1 ▪ Aktí Tóti Xatzí, Égina Town ▪ 22970 23594 ▪ €

Nostalgic images adorn the walls of this great taverna overlooking the bay. Dine inside or on the terrace. The menu is classic Greek with a contemporary twist.

(6) Remetzo
MAP K2 ▪ Pérdika ▪ 22970 61658 ▪ €€

The delicious menu at this harbourside restaurant features octopus to start, followed by the speciality, lobster with pasta. Fine wine completes the meal *(see p64)*.

(7) Flisvos
MAP K1 ▪ Leofóros Kazantzáki, Égina Town ▪ 22970 26459 ▪ €

Known for its shellfish and fresh fish dishes, this restaurant lies on the water's edge, looking out over the bay. Its selection of wines and beers feature local brews.

(8) O Skotadis Restaurant
MAP K1 ▪ Paralía Egínas, Égina Town ▪ 22970 24014 ▪ €

This seafront taverna offers an extensive selection of salads and starters, followed by meats and seafood such as anchovies, squid and sardines. Watch the boats in the bay as you dine.

(9) Plaza Restaurant
MAP K1 ▪ Kazantzaki 4, Égina Town ▪ 22975 00712 ▪ €

Enjoy fantastic sunsets while dining at this restaurant right by the beach. There are seafood specials and a wide range of organic wines.

(10) Costantonia
MAP L1 ▪ Agía Marína ▪ 22970 32192 ▪ €€

A lively restaurant and bar, Costantonia offers breakfast, coffee, cocktails and evening dining delicacies such as grills with homemade dips and delicious *mezédes*.

See map on pp144–5

Streetsmart

An archetypal street scene on the
Greek Islands

Getting To and Around the Greek Islands

Arriving by Air

The Greek Islands are served directly by many European airlines and by several intercontinental companies via domestic connections from the **Elefthérios Venizélos Airport (Athens)**.

Most scheduled flights arrive at the islands' main airports, such as Crete's **Ioannis Daskalogiannis (Chaniá)** and **Nikos Kazantzakis (Irákleio)**, **Dionysios Solomós (Zákynthos)**, **Ioannis Kapodistrias (Corfu)**, **Diagoras (Rhodes)**, **Aristarchos (Sámos)**, **Hippocrates (Kos)**, **Ifestos (Límnos)** and **Mytilene Odysseas Elytis (Lésvos)**.

Domestic flights link the islands; tickets can be purchased online. The two main operators are **Olympic Air** and **Aegean Airlines**.

Arriving by Sea

The island groups are well served by passenger ships, ferries and fast hydrofoils – many are luxurious with amenities. The large port of Piraeus in Athens has dozens of boats leaving for Crete, the Cyclades, the Dodecanese, the North and South Aegean, and the Saronic Gulf every day. Rafina port in Athens also serves some of the islands. Other ports such as Igoumenitsa, Pátra and Kyllini serve the Ionians. Tickets and information on sailing times can be found online. Be aware that sailings can often be cancelled without notice if the weather is bad. The main sea operators are **Hellenic Seaways**, **Blue Star Ferries**, **Aegean Speed Lines**, **SeaJets** and **Dodekanisos Seaways**, along with a few smaller ones.

Yachtsmen have a wide choice of marinas and anchorages where they can come ashore. Among them are the large Gouvia Marina and Limin Kérkyra Marina on Corfu, the well-equipped Kos Marina, Ágios Nikólaos at Lassithi on Crete, the Marina Mandráki on Rhodes and the marinas at Skiáthos and Sýros.

The Greek Islands appear on many cruise-ship itineraries. These floating hotels can often be seen anchored a little way off Corfu Town's coastline or at the harbour of Zákynthos. Other popular ports of call include Rhodes and Pátmos in the Dodecanese, Santoríni and Mýkonos in the Cyclades, and Crete.

Arriving by Road

Other than Évvia and Lefkáda, which are linked to the mainland by road bridges, all the other islands are accessed only by air or sea. It is possible to island-hop around some of the groups if you are travelling by car as the ferries are large and can accommodate various modes of transport from cycles to goods lorries.

Getting Around by Bus

Most bus services are operated by **KTEL**'s regional divisions, such as those in Corfu and Crete, or small, privately owned local companies, with timetable regularity varying between the islands. Bus stations where you can check timetables and buy tickets can usually be found in the islands' main town harbours. A journey can feel like stepping back in time as many vehicles are old, especially those on the remotest islands. There are intercity buses linking major islands' towns and capitals to Athens, which include the cost of the ferry.

Tour operators, and hotels, especially those in holiday hotspots, organize excursions that take in major sights. Excursions usually include a commentary giving historical and practical information.

Getting Around by Taxi

It is generally easy to find a taxi on any of the islands. The main towns have dedicated taxi ranks and even the smallest villages have a vehicle available to hire. Taxis are metered and inexpensive, but if hiring a taxi for a day's sightseeing it is a good idea to negotiate a fixed price beforehand. Taxi drivers are obliged by law to issue a receipt for the total cost of a journey on request. It must clearly state the driver's name and the vehicle's registration number. Taking a taxi tour is a good way to explore as the driver will almost certainly have lots to tell you about local life.

Getting Around by Car

All the airports, resorts and many smaller towns have car hire company offices, including **Sixt**, **Europcar** and **Budget** in Corfu and Crete. You can hire a 4WD if planning to explore off-road, or a minibus if travelling with a group. Family cars are also available. A full driving licence held for at least a year is required, and US licence holders must have an International Driving Permit. Driving is on the right, with priority from the right at junctions. Observe speed limits and be especially careful when driving in mountainous areas or villages as roadways can be narrow and uneven. Highways tend to be in good order. Children under 12 must not travel in the front.

Getting Around by Bicycle

Cycling is an enjoyable way to explore the islands and most of the resorts have places where you can hire a cycle or a mountain bike for a day, a couple of days or longer. Local companies include **Corfu Mountain Bikes**, **Cycling Creta**, **SD Bikes**, **Zakynthos Cycling Centre**, **Kos Mountainbike Activities** and **Rent a Bike Kefalonia**. Tourist offices on the main islands publish brochures with suggested cycle routes. Some are short, leisurely rides, while others are designed for more experienced cyclists and might cover long distances or more challenging terrain. Take a fully charged mobile phone and ensure you have plenty of water with you when cycling to avoid dehydration.

DIRECTORY

ARRIVING BY AIR

Aegean Airlines
W aegeanair.com

Aristarchos (Sámos)
C 22730 87800

Diagoras (Rhodes)
W rho-airport.gr

Dionysios Solomós (Zákynthos)
C 26950 29500

Elefthérios Venizélos Airport (Athens)
W aia.gr

Hippocrates (Kos)
C 22420 56000

Ifestos (Límnos)
C 22540 29400

Ioannis Daskalogiannis (Chaniá)
W chania-airport.com

Ioannis Kapodistrias (Corfu)
W cfu-airport.gr

Mytilene Odysseas Elytis (Lésvos)
C 22510 38700

Nikos Kazantzakis (Irákleio)
W heraklion-airport.info

Olympic Air
W olympicair.com

ARRIVING BY SEA

Aegean Speed Lines
W aegeanspeedlines.gr

Blue Star Ferries
W bluestarferries.gr

Dodekanisos Seaways
W 12ne.gr

Hellenic Seaways
W hellenicseaways.gr

SeaJets
W seajets.gr

GETTING AROUND BY BUS

KTEL
W ktelbus.com

GETTING AROUND BY CAR

Budget
W budget.gr
Corfu Office: **MAP B5**;
Eth Lefkimmis 29,
Corfu Town
Crete Office: **MAP E6**;
Irákleio Airport

Europcar
W europcar.com
Corfu Office: **MAP B5**;
K Georgaki, Corfu Town
Crete Office: **MAP E6**;
Ikarous 97, Irákleio

Sixt
W sixt.com
Corfu Office: **MAP B5**; El
Venizélou 46a, Corfu Town
Crete Office: **MAP E6**;
Kazantzakis Airport,
Irákleio

GETTING AROUND BY BICYCLE

Corfu Mountain Bikes
W corfumountainbikes.com

Cycling Creta
W cyclingcreta.gr

Kos Mountainbike Activities
W kosbikeactivities.com

Rent a Bike Kefalonia
W rentabikekefalonia.gr

SD Bikes
W sdbikes.gr

Zakynthos Cycling Centre
W podilatadiko.com

Practical Information

Passports and Visas

Visitors from outside the European Economic Area (EEA), European Union (EU) and Switzerland need a valid passport to travel to Greece, as do UK visitors; most other EU nationals require only a valid identity card. Citizens of Canada, the USA, Australia and New Zealand do not need visas for stays of up to 90 days as long as their passport is valid for 6 months beyond the date of entry. For longer stays, a visa is necessary and must be obtained in advance from the Greek embassy. For details and useful information, check the website of the Greek **Ministry of Foreign Affairs**.

Most countries have consular representation in Athens, including the **USA**, **Canada**, the **UK** and **Australia**. The **New Zealand** embassy in Rome is accredited to Greece, and there is also a consulate in Athens.

Customs and Immigration

Duty-paid goods such as alcohol and perfumes for personal use are no longer subject to official limits within the EU. There is no restriction on the amount of money EU citizens can bring for holiday use when visiting Greece. However, large sums should be declared on entry.

Visitors from a non-EU country entering Greece and the Greek Islands by air or sea may bring in 200 cigarettes or 50 cigars, one litre of spirits, four litres of wine and 16 litres of beer. If entering the country by land then tobacco products must not exceed 40 cigarettes or 10 cigars. The amount of money visitors from non-EU countries may bring in to Greece must not exceed €10,000 or the equivalent value in another currency.

Importing perishable food items from non-EU countries is strictly prohibited, as is the removal of archaeological artifacts from the country.

Travel Safety Advice

Visitors can get up-to-date travel safety information from the **UK Foreign and Commonwealth Office**, the **US Department of State** and the **Australian Department of Foreign Affairs and Trade**.

Travel Insurance

Getting travel insurance is advisable for everyone travelling to the Greek Islands, and should cover theft, loss, medical problems, and any delays to travel arrangements.

Health

No immunizations are required when visiting the Greek Islands. The islands are also unaffected by most dangerous infectious diseases.

The sun can be very hot in the summer so be sure to wear sunscreen and a hat when outdoors, and drink plenty of water to avoid dehydration. It is generally safe to drink tap water, but it may be wise to buy bottled water from supermarkets or kiosks to avoid the risk of an upset stomach through unfamiliar chemical content, or simply for convenience when out sightseeing. Take water with you when visiting archaeological sites and outdoor attractions as drinking water isn't always available on site and can be expensive. If you do get dehydrated seek medical help.

Mosquitoes do not carry serious diseases in the region but they can leave a nasty bite and transfer germs. Some people may even get a reaction to a mosquito bite, in which case a pharmacy should be able to advise. Wear repellent, especially after dark or when near water, and invest a plug-in deterrent.

The medical facilities provided by the Greek National Health Service are generally good throughout the islands, although facilities on the smaller islands can be limited. In the larger towns, hospitals are modern with accident and emergency departments, such as those at the large **Corfu General Hospital Iagia Eirini** or the **Chaniá General Hospital St George** on Crete.

In a medical emergency, phone for an **ambulance**. Ambulance response times are generally good, but as few roads have names you should be prepared to give details of local landmarks when giving directions. Air ambulances are available to fly serious emergency cases to the state-of-the-art hospitals in Athens.

EU citizens are entitled to free emergency medical care because of reciprocal arrangements that are in place with Greece and its islands, but they will need to show a valid passport and European Health Insurance Card (EHIC) for registration at the hospital. Visitors should obtain an EHIC before travelling, which is available from the **Department of Health**.

The majority of the major towns throughout the Greek Islands have private medical practices. Doctors' private fees are payable immediately on treatment and a receipt is provided for holiday insurance purposes. You should find your hotel can recommend local doctors who can assist with minor medical problems.

For mild ailments you can consult pharmacists, who are highly trained. Alongside advising the patient on the problem, they can prescribe and dispense certain medicines. Pharmacies are easily identified by a large, usually illuminated, green cross and stock a comprehensive range of medication. If you take prescription medication it is wise to pack sufficient for the length of your stay as local pharmacies may not stock the exact same medication as your own doctor has prescribed. You should make sure you carry an accompanying doctor's note with your prescribed medication in case an official questions the nature of your drugs. Call the out-of-hours **Pharmacies Hotline** for detailed information on the opening times of local pharmacies.

In the event of a dental emergency, hotels can also recommend dentists. Dental practices are run on a private basis with emergency treatment fees payable immediately. Receipts are given for insurance purposes.

Personal Security

The Greek Islands have a low crime rate, but thefts from tourists do occur. Take sensible precautions, including locking rental cars and hotel rooms, and keeping passports, tickets and spare cash in hotel safes. If you have things stolen, contact the **Tourist Police**; they speak other foreign languages, including English. In emergencies, call the relevant **Police** or **Fire** numbers.

DIRECTORY

PASSPORTS AND VISAS

Australia
MAP D3; Corner of Kifisias & Alexandras Ave, Ambelokipi, Athens
21087 04000
greece.embassy.gov.au

Canada
MAP D3; 48 Ethnikis Antistaseos St, Athens
21072 73400
greece.gc.ca

New Zealand
MAP D3; 76 Leoforos Kifisias, Athens
21069 24136
mfat.govt.nz

Ministry of Foreign Affairs
mfa.gr/en

UK
MAP D3; 1 Ploutarchou Street, Athens
21072 72600
ukingreece.fco.gov.uk

USA
MAP D3; 91 Vasilis Sofia Avenue, Athens
21072 12951
gr.usembassy.gov

TRAVEL SAFETY ADVICE

Australian Department of Foreign Affairs and Trade
dfat.gov.au
smartraveller.gov.au

UK Foreign and Commonwealth Office
gov.uk/foreign-travel-advice

US Department of State
travel.state.gov

HEALTH

Ambulance
166

Chaniá General Hospital St George
MAP D6; Moyrnies
28210 22000

Corfu General Hospital Iagia Eirini
MAP B5; Kontokali
26613 60400

Department of Health
nhs.uk/ehic

Pharmacies Hotline
14944

PERSONAL SECURITY

Fire
199

Police
100

Tourist Police
171

Currency and Banking

The official currency of the Greek Islands is the euro (€), divided into 100 cents. However, sterling is also readily accepted at hotels and restaurants at a good rate. Banknotes come in denominations of €5, €10, €20, €50, €100, €200 and €500 (the latter is rarely seen), and coins are 1c, 2c, 5c, 10c, 20c and 50c *(lepta)*, and €1 and €2. Visitors coming from the EU may bring in an unlimited amount of foreign currency, but large sums should be declared on arrival at customs.

Changing money is straightforward, even in smaller towns, and can be done over the counter at a bank. Opening hours are 8am–2pm weekdays and until 1:30pm Friday. Banks, such as the **Piraeus Bank**, **Alpha Bank** and the **National Bank of Greece**, which have branches throughout the islands, are generally closed in the afternoons and on Saturday, although in major tourist areas some stay open later, especially during summer. The bank may ask to see proof of identity.

Smaller towns and even villages are catching up by having their own ATMs, but these machines are usually only found in the main towns and in tourist areas. Operating round the clock and dispensing local currency, most ATMs accept all major cards.

All major credit cards are widely accepted by hotels, larger restaurants and shops in the main towns, but cash is often preferred in smaller establishments. It is rare to find a village store and taverna willing to accept anything other than cash.

Telephone and Internet

Mobile phone usage is widespread in Greece and visitors who bring their own phone are unlikely to experience any problems when calling home or a local number. If you have an international GSM-equipped phone, check with your service provider at home if global roaming is available to you and how much it will cost. Alternatively, consider buying a SIM card locally. Mobile providers include **Vodafone** and **Cosmote**.

The country code for Greece is 30. Telephone numbers start with a service code (for example: 2 indicates a landline, 6 for a mobile number and 9 a premium rate number), followed by an area code and then the eight-digit private subscriber number. Athens' code is 21, so if dialling the city from outside Greece the number would be +30 210 0000000. Within Greece, it would be 210 0000000. Most of the islands or island groups have their own codes: Évvia is 222, the Dodecanese 224, Lésbos and Límnos 225, Híos and Sámos 227, the Cyclades 228, some of the Argo-Saronic islands 229, Corfu 266, Kefalloniá 267, Zákynthos 269, and Crete 281–289. To phone home from Greece dial 00 for the international operator, followed by the country code. For example, 44 for the UK minus the first 0 of the private number, 1 for the USA, 353 for Ireland, 61 for Australia and 64 for New Zealand.

The fast-moving world of technology has not escaped the islands and although not all homes have an internet connection it is becoming more commonplace by the day. Wi-Fi is widespread and available in nearly every hotel, bar and restaurant, normally for free or for just a small fee.

Postal Services

Post offices operated by **Hellenic Post SA (ELTA)** can be found in all the major towns and most of the villages, and are easily identified by the yellow signs that read *tachydromeia* (post office). Letter boxes are painted yellow. Opening times of post offices may vary between the islands, especially in the smaller villages, but are generally around 7:30am–2pm. Larger post offices often stay open later and on Saturday, especially in holiday resorts.

Visitors can also post letters at the reception desk of most hotels. Letters and postcards sent to European countries will usually arrive at their destination in around five days, but may take longer if posted in a remote village where post will be sent to the nearest sorting office before continuing its journey.

TV, Radio and Newspapers

Greece's national state broadcaster is **Ellinikí Radiofonía Tileórasi (ERT)**. ERT's transmissions can be picked up around the islands, plus there are numerous privately owned stations. Most broadcast in Greek with scheduled programmes in English. The quality of transmission and production is mixed. Hotels tend to have satellite television installed with international news and lifestyle channels.

British and European newspapers are usually found in kiosks, supermarkets and hotel foyers the day after publication, although the markup on them can be substantial. International lifestyle and business magazines are generally available too, again with a hefty markup. English and Greek language local publications are also available, which tend to focus on current news and upcoming events. Many of the islands or groups have their own publications, such as the *Crete Gazette*. Also the *Greek Reporter* carries information on the islands and *Ekathimerini* publishes news online from all over Greece.

Opening Hours

Shops are generally open 9am–2pm and 5:30–9pm on weekdays, and close at 2pm on Wednesday and Saturday. Smaller shops may also have a long lunch break, although with Greece's recent financial woes the pressure to trade has increased and many businesses now stay open all day. Banks are open 8am–2:30pm weekdays except until 2pm on Friday, although those

in major tourist areas may stay open later.

Restaurants usually open daily at 11am–4pm for lunch, and 7pm–midnight for dinner, with one day a week when they are closed; the day can vary.

The opening times of museums and major sights, such as the Palace of Knossos on Crete or the Temple of Aphaia on Égina, can vary enormously, but tend to be 8am–3pm daily except Monday, and often until 8pm in summer.

Time Difference

Greece is normally 2 hours ahead of GMT. Daylight saving time applies in the summer months with clocks going forward 1 hour on the last weekend of March, and back 1 hour on the last weekend of October.

Electrical Appliances

The standard current is 230V/50Hz and plugs are of the two round pin variety, although in some cases they have three round pins. If travelling from the UK you will need an adaptor for your appliances. A transformer will be required for some American appliances. Adaptors and transformers are not always available in Greece, and especially on the islands, so make sure you pack what you need.

Weather

The Greek Islands usually have hot summers, milder autumns and springs, and cool winters

when activity holidays are popular. Temperatures rarely fall below 10° C (50° F) on the islands' coasts, but snow often falls in the mountainous regions. December and February are the coolest, wettest months, while July and August are the hottest, with temperatures averaging around 30° C (86° F). The sun can be strong in the summer, even on overcast days, and it is advisable to use sunscreen, wear good sunglasses and a hat, and drink plenty of water.

Travellers with Specific Needs

Unfortunately, there are very few public buildings, shops, museums and archaeological sites that have wheelchair ramps so access can be difficult for wheelchair users. Many museums are in older buildings without lifts.

Most hotels and some restaurants are starting to introduce ramps and bathroom facilities that are suitable for accommodating wheelchairs. The tourist authority publishes an informative leaflet entitled Accessible Greece, which can be obtained from all tourist information offices.

Few places around the islands provide Braille or audio guides adapted for visually impaired visitors, nor induction loop devices for those with hearing difficulties.

There are specialist companies, such as **Disabled Holidays**, that organize package holidays in Greece for travellers with special needs. The Athens-based **AccessibleTravel** provides advice on accessibility on its website.

Travelling with Children

If travelling with children, care should be taken to avoid sunburn and accidents in strong sea currents. Supermarkets and pharmacies generally have supplies of baby food, children's medicines and nappies, although take any special dietary needs with you as the same brands may not be available.

Sources of Information

The **Greek National Tourism Organization (GNTO)** has offices in the UK, USA, Germany and many other countries. Full address details can be found on the respective websites or on the GNTO's own website. The GNTO's headquarters is in Athens. There are GNTO offices throughout the islands, including on **Corfu**, **Crete**, **Kos**, **Lefkáda**, **Rhodes**, **Santoríni** and **Sýros**. A number of helpful blogs about the region are also springing up; one of the best is **Greeka**.

Shopping

The Greek Islands' best shopping streets tend to be off the seafront of the larger islands' towns and around the harbourside of the smaller islands. Many are specialist arts and crafts shops where handmade items can be purchased to take home as a souvenir (see p66–7). In Rhodes Old Town, for instance, head for bustling Ippokratous Square where you'll find ceramic pieces, original paintings, copperware, jewellery shops and trendy fashions. In contrast, on an island like Sýros you'll find artisans' workshops in Ermoúpoli harbour.

Most towns in the Greek Islands have a weekly street market, known as the laïkí agora, where you'll be able to purchase a wide range of fresh fruit and vegetables, meat and fish, herbs, a whole host of household items, crafts, children's toys and cheaper clothes.

Dining

The Greek Islands offer a choice of places to eat from upmarket restaurants serving gourmet cuisine to smaller eateries. Look out for family-owned tavernas where local dishes will almost certainly be freshly prepared and cooked. Every village, however small, will have at least one café, known as a kafeneía. While traditionally they serve only coffee, more now serve sandwiches and light meals, and are more akin to a snack bar. American-style hamburger restaurants, pizza parlours and ice-cream shops have made their mark on the Greek Islands too, and can be found in most major towns. The local equivalent is a souvlatzídiko, which sells delicious souvláki made from chunks of pork or chicken chargrilled with herbs.

The Greek diet largely comprises meat, fish, vegetables and fruit. Dairy products like yoghurt are often used in the preparation, as are nuts. Visitors with nut allergies, diabetes or lactose or gluten intolerances should exercise care. Some restaurants will cater for special diets. Dishes like vegetable moussakás appear on many menus, although few restaurants are exclusively vegetarian or vegan.

A range of excellent beers are brewed in Greece, the most famous being Mythos, while popular spirits include tsípouro made from grapes and the aniseed-flavoured *ouzo*. There are local versions of white rum, brandy and other spirits. The traditional wine of Greece is *retsina*.

A service charge is added to the price of meals on the menu, and should be clearly stated. Having said that, if the service you received was good then you might like to consider rounding up the bill or adding a 10 per cent tip.

Most restaurants and tavernas will welcome families with children. Nice touches such as special menus, colouring books with crayons and a wide range of soft drinks are provided.

Where to Stay

The Greek Islands' hotels and accommodation are graded from five star to one star, although in practice there are very few two-star hotels or less on the islands. In general, it is best to avoid hotels with fewer than three stars as these tend to have few facilities and are likely to be in a poor location. There is also a deluxe category for luxury hotels, which have superior facilities like a spa and gourmet restaurants. Most hotels have different dining options depending on the package booked. Visitors might also like the flexibility of booking only on a B&B basis or prefer half board, full board or All Inclusive.

Self-catering rooms, apartments and holiday villas are graded A to C. In all cases, the emphasis is on the facilities provided, such as a communal swimming pool, a maid service and the number of bedrooms. Ranging from large, luxurious villas to small rooms, this type of accommodation is usually advertised privately or through a specialist company.

Rates and Booking

Rates can fluctuate a bit, depending on supply and demand. Prices generally peak at the height of the summer season in July and August, but rates can also rise in accordance with local festivals and events. It is always best to do your research and book your accommodation several months in advance during popular times. Room prices will almost always include your breakfast.

In most cases, making a booking is straightforward by telephone and email directly with the hotel, or through a specialist online company like **HotelsCombined**, **Booking.com**, **Trivago**, **Expedia**, **Lastminute.com** or **Agoda**. You can also rent apartments and private rooms directly from the owner via websites such as **Airbnb**, **HomeAway** and **FlipKey**.

DIRECTORY

TRAVELLERS WITH SPECIFIC NEEDS

AccessibleTravel
w accessibletravel.gr

Disabled Holidays
w disabledholidays.com

SOURCES OF INFORMATION

Greeka
w greeka.com

Greek National Tourism Organization (GNTO)
MAP D3; 7 Tsoha Street, Ampelokipoi, Athens; 21087 07000
w visitgreece.gr

GNTO Corfu
MAP B5; Evangelistrias 4, Corfu Town
📞 26610 37520

GNTO Crete
MAP E6; Papa Alexandrou 16, Irákleio
📞 22810 246106

GNTO Kos
MAP Y1; Artemisias 2
📞 224 202 9910

GNTO Lefkáda
MAP H1; Lefkáda Marina
📞 26450 25292

GNTO Rhodes
MAP U4; Papagou and Archiepiskopou Makariou
📞 22410 44330

GNTO Santoríni
MAP V2; Fira
📞 22860 27199

GNTO Sýros
MAP M6; Thimaton Sperheiou 11
📞 22810 86725

WHERE TO STAY

Agoda
w agoda.com

Airbnb
w airbnb.co.uk

Booking.com
w booking.com

Expedia
w expedia.com

FlipKey
w flipkey.com

HomeAway
w homeaway.co.uk

HotelsCombined
w hotelscombined.com

Lastminute.com
w lastminute.com

Trivago
w trivago.com

Places to Stay

PRICE CATEGORIES

For a standard, double room per night (with breakfast if included), taxes and extra charges.

€ under €100   €€ €100–€300   €€€ over €300

Luxury and Mid-Range Hotels in Corfu

Bella Venezia

MAP B5 ■ Zambéli 4, Corfu Town ■ 26610 46500 ■ www.bella veneziahotel.com ■ €€

A town-centre hotel in an apricot-coloured Neo-Classical mansion, Bella Venezia offers elegance and seclusion. Dining in the courtyard garden is a memorable experience.

Corfu Palace

MAP B5 ■ Leoforos Dimokratias 2, Corfu Town ■ 26610 39485 ■ www. corfupalace.com ■ €€

Set in subtropical gardens overlooking the Garítsa bay, the deluxe Corfu Palace has well-equipped rooms with marble baths and sea views, pools, restaurants such as the respected Scheria, and a spa offering seawater treatments.

Family Life Kerkyra Golf

MAP B5 ■ Alykés Potamoú ■ 26610 24030 ■ www. familylifekerkyragolf.com ■ €€

Ideal for an activity-filled holiday, the Family Life Kerkyra Golf has tennis courts, restaurants and a nearby 18-hole golf course. To keep the children happy, there is a family entertainment programme, plus a mini club and playground.

Marbella Corfu

MAP B6 ■ Ágios Ioánnis ■ 26610 71183 ■ www. marbella.gr ■ €€

The Marbella Corfu is an eco-friendly, five-star hotel with well-appointed rooms. The complex boasts a concert hall, spa, fitness centre and restaurants, including La Terrazza, which has great views looking out over the beach.

Pelecas Country Club

MAP A5 ■ Pélekas ■ 26610 52918 ■ www. country-club.gr ■ €€

The restoration of a former mansion and its outbuildings, including the stables and summer house, created this luxurious club. Occupying a private estate from the 18th century, it offers glamourous suites and studios, a swimming pool, tennis courts and a dining hall.

Grecotel Corfu Imperial

MAP B5 ■ Komméno ■ 26610 88400 ■ www. corfuimperial.com ■ €€€

Gourmet restaurants, elegant rooms, a spa and berthing for yachts are just some of the super facilities offered at the five-star Grecotel Corfu Imperial. The hotel stands on a stunning peninsula overlooking the bay of Corfu, and has Italianate-style gardens and decor.

Budget Stays and Camping in Corfu

Anemona Studios

MAP A5 ■ Paleokastrítsa ■ 26630 41101 ■ €

A building consisting of ten apartments, some with private balconies, the Anemona Studios lies on Paleokastrítsa resort's periphery, with amenities close by. Guests can use the nearby Phivos Studios' swimming pool and bar.

Casa Lucia

MAP A5 ■ Sgómbou ■ 26610 91419 ■ www. casa-lucia-corfu.com ■ €

Set in the countryside, Casa Lucia comprises former olive-press buildings that have been remodelled into self-catering cottages. Many date to the Venetian period and feature vaulted ceilings. Painting, yoga and T'ai Chi classes are offered to guests.

Dionysus Camping Village

MAP B5 ■ Danílas Bay, Dasiá ■ 26610 91417 ■ www.dionysuscamping.gr ■ €

One of Corfu's oldest campsites, the Dionysus is set amid gardens and forest. It welcomes caravans and camper vans, and has bungalows and tents for hire. On-site are shower blocks, a supermarket and a restaurant.

Fundana Hotel

MAP A5 ■ Odysséos 1, Paleokastrítsa ■ 26630 22532 ■ www.fundana villas.com ■ €

Full of character, the Fundana Hotel is housed in a lovely pink-washed,

17th-century Venetian mansion standing in terraced gardens full of hibiscus and olive trees. The stylish rooms and suites are welcoming. There are sun terraces around the pool and a small terrace for dining.

Hotel Zafiris

MAP A4 ▪ Melítsa Perouládes ▪ 26630 95086 ▪ www.zafirishotel.gr ▪ €
This modern two-star hotel comprises studios and apartments laid out around gardens and walkways. Amenities include the Waterfalls restaurant, which uses vegetables from its own organic garden, and the Saxophone cocktail bar.

Paleokastrítsa Camping

MAP A5 ▪ Paleokastrítsa ▪ 26630 41204 ▪ www.campingpaleokastritsa.com ▪ €
Shower blocks, electricity, a children's play area, and the use of a swimming pool in its sister holiday complex, are just some of the facilities here. Ideally positioned in shady olive groves, this campsite is open to caravans, camper vans and visitors with tents.

Thekli-Clara Studios

MAP B2 ▪ Gäios, Paxí ▪ 26620 32313 ▪ www.theklis-studios.com ▪ €
This attractive group of stylish and well-equipped whitewashed studios is in the centre of Gäios. Many units have balconies with wonderful views that are ideal for relaxing and watching the local life. There are harbour-side taverns close by.

Luxury and Mid-Range Hotels in the Ionians

Hotel Palatino, Zákynthos

MAP H4 ▪ Kolokotróni 10, Zákynthos Town ▪ 26950 27780 ▪ www.palatinohotel.gr ▪ €
Located on the seafront, this four-star hotel is ideal for a holiday or business stay. Trinity restaurant has both Greek and international fare on the menu, and the Palatino Café & Cocktail Bar has a welcoming atmosphere. Amenities include tennis courts.

Armonia, Lefkáda

MAP H1 ▪ Megálo Avláki, Nydrí ▪ 69375 10586 ▪ www.armonianakas.blogspot.com ▪ €€
A modern hotel with traditional architecture, Armonia offers well-furnished rooms, some with vaulted ceilings and sea views. The restaurant's terrace looks over the bay. Other amenities include a pool and a cocktail bar.

Astir Palace, Zákynthos

MAP H4 ▪ Laganás Bay ▪ 26950 53300 ▪ www.astirhotels.gr ▪ €€
This modern resort-style hotel has great gardens, a swimming pool and a children's play area. The 120 air-conditioned rooms have balconies overlooking the sea. The hotel restaurant serves international cuisine.

Caravel Zante, Zákynthos

MAP H4 ▪ Plános, Tsiliví ▪ 26950 45261 ▪ www.caravelzante.gr ▪ €€
This medium-sized hotel enjoys an idyllic location, surrounded by olive groves and overlooking the beach. It has a Fun Park on-site, along with restaurants and a theatre. Rooms have sea views.

Cephalonia Palace, Kefalloniá

MAP G3 ▪ Xi Lixourioú ▪ 267 10 93112 ▪ www.cephaloniapalace.com.gr ▪ €€
There is a choice of room types, including family rooms, available at this modern four-star resort. Guests can relax around its freshwater pool, lounge on its terraces or try horse riding, sailing or scuba diving. Children have their own pool and playground.

Grand Nefeli Hotel, Lefkáda

MAP G1 ▪ Póndi, Vasilikí ▪ 26450 31378 ▪ www.grandnefeli.com ▪ €€
A four-star hotel, Grand Nefeli, located right on the beach, offers an array of facilities, including a children's playground, swimming pool, terraces and pretty gardens. It has its own windsurfing tutors as well. Rooms feature climate control.

Ionian Blue, Lefkáda

MAP H1 ▪ Nikiána ▪ 26450 29029 ▪ www.ionianblue.gr ▪ €€
Set on a hillside with panoramic views of the breath-taking sea, the Ionian Blue is one of the island's foremost hotels. Luxurious and elegant, it provides all amenities including a spa, private swimming pools, restaurants and an intimate gourmet restaurant.

Mareblue Apostolata Resort & Spa, Kefaloniá

MAP H3 ▪ Skála
▪ 26710 83581 ▪ €€
This four-star hotel is among the finest on the island. It has a spa and several restaurants, including the Sea Pearl and Sunrise, serving buffets and à la carte cuisine. Tennis and volleyball are available.

Mediterranean Beach Resort, Zákynthos

MAP H4 ▪ Seafront, Laganás Bay ▪ 26950 55230 ▪ www.med beach.gr ▪ €€
Situated on the long sandy beach of Laganás, this resort breaks away from the mould of other luxury resorts and resembles a typical Ionian village. Each one of the buildings is named after a Greek colony. There is an excellent restaurant and spa on site *(see p46)*.

Paxos Beach Hotel, Paxi

MAP B2 ▪ Gäios
▪ 26620 32211 ▪ www. paxosbeachhotel.gr ▪ €€
Set in a house made of local stone, with bunga-lows dotted amongst terraces of olive trees, this hotel overlooks a pretty bay and has a private beach. There is a swim-ming pool, a playground and the restaurant serves excellent local dishes.

Trapezaki Bay Hotel, Kefaloniá

MAP G3 ▪ Trapezaki, Kefaloniá ▪ 26710 31502 ▪ www.trapezakibay hotel.com ▪ €€
Well-positioned on a hilltop, this adults-only hotel offers dramatic

views, tranquillity and peace. All of the 33 rooms here are comfortably furnished, and the restaurants offer some of the best Greek food to be found on the island.

Budget Stays and Apartments in the Ionians

Allegro Hotel, Kefaloniá

MAP G3 ▪ A Hoidá 2, Argostóli ▪ 26710 22268 ▪ €
Conveniently located near Argostóli's central square, the Allegro is ideal for guests wishing to explore the capital. The air-conditioned rooms are equipped with good music systems and there is an on-site taverna with great bay views.

Aris Villas, Lefkáda

MAP H1 ▪ Ligia
▪ 26450 71260 ▪ €
Just a few minutes' walk from the beach, this small hotel of 12 apartments and studios is built in a traditional style with whitewashed walls and set in lush gardens and olive groves. There are stunning panoramic views available from the terrace.

Astoria Hotel, Zákynthos

MAP H4 ▪ Alykés bay
▪ 26950 83533 ▪ www. astoriazante.com ▪ €
A long-established family-run hotel, the Astoria is set right next to the beach. It provides stylish guest rooms, plus there is a lounge bar, a shop and a restaurant that serves traditional Greek dishes using fresh local produce.

Bel Air Hotel, Lefkáda

MAP H1 ▪ Nydrí
▪ 26450 92125 ▪ www. hotel-belair.gr ▪ €
An attractive terracotta and cream complex, the Bel Air Hotel is located minutes from the beach. It has 33 air-conditioned apartments, swimming and spa pools, plus a children's pool. It also has bars and an alfresco dining terrace.

Grivas Gerasimos Apartments, Itháki

MAP H2 ▪ Vathý bay
▪ 26740 33328 ▪ €
A modern building overlooking Vathý bay, the Grivas Gerasimos comprises two- and three-bedroom apartments all with a balcony or terrace. Air-conditioning and an equipped kitchen are among the facilities, and a daily maid service is provided. Nearby there is a good selection of tavernas and shops.

Leedas Village, Zákynthos

MAP H4 ▪ Lithákia
▪ 26950 51305 ▪ www. leedas-village.com ▪ €
A super collection of villas built in traditional style using local stone, this self-catering complex is good value for money. The decor is stylish. Features include stone walls and timber ceilings, a garden, barbecue, children's play area and communal pool.

Linardos Apart-ments, Kefaloniá

MAP G2 ▪ Ássos ▪ 26740 51563 ▪ www.linardos apartments.gr ▪ €
This whitewashed hotel is an integral part of the

scenery around the attractive bay at Ássos. Stylishly decorated, the air-conditioned studios and a two-storey apartment (designed for up to six people) come complete with an equipped mini-kitchen and a daily maid service.

Odyssey Apartments, Itháki

MAP H2 ▪ Vathý ▪ 26740 33400 ▪ www.odyssey apartments.gr ▪ €
This complex is a real find for travellers on a budget. Situated minutes from the beach and town centre, it comprises studios and apartments with balconies and kitchens, all decorated in a fresh colour scheme. Breakfast is served on the terrace with views over a pool.

Ostria Hotel, Lefkáda

MAP G1 ▪ Ágios Nikítas ▪ 26450 97483 ▪ €
Decorated in a traditional Greek style, this small pension-style hotel is set on a hillside overlooking the bay. Each of its 12 rooms has a balcony. The owners are keen chefs and use local ingredients to prepare the classic Greek cuisine served in its restaurant.

Luxury and Mid-Range Hotels in the Cyclades

Aphrodite Paradise Beach, Mýkonos

MAP P6 ▪ Kalafáti beach ▪ 22890 71367 ▪ www. aphrodite-mykonos.com ▪ €€
This lively resort has its own nightclub where events such as retro discos and Greek music nights take place. Other

facilities include bars and a restaurant, a boutique and a mini-market.

Archipelagos Resort, Páros

MAP E4 ▪ Agía Iríni ▪ 22840 24176 ▪ www. archipelagosresort.com ▪ €€
One of the finest boutiques in the Cyclades, Archipelagos offers 30 luxurious suites as well as three family villas. Guests can participate in a wide range of activities, including watersports, mountain biking and excursions (see p46).

Astir of Paros, Páros

MAP E4 ▪ Náoussa ▪ 22840 51976 ▪ www. astirofparos.gr ▪ €€
Standing in a large garden full of date palm trees and hibiscus, this luxury seafront complex resembles a traditional Cycladic village. Chinese and Mediterranean-themed restaurants, an art gallery, pools and a fitness suite are just some of the facilities.

Dolphin Bay, Sýros

MAP M6 ▪ Galissás ▪ 22810 42924 ▪ www. dolphin-bay.gr ▪ €€
This four-star hotel has 140 rooms with a range of amenities and views of the bay. Meals are served in a restaurant and traditional taverna. In the grounds is a half Olympic-size pool, and a children's playground.

Faros Hotel, Sýros

MAP N6 ▪ Azólimnos ▪ 22810 61661 ▪ www. faros-hotel.com ▪ €€
Housed in blue and white buildings, this four-star hotel comprises studios,

apartments and suites, all with sea views, surrounding a swimming pool. A mini shopping mall, hairdresser and fitness suite are available for guests on site.

Harmony Boutique Hotel, Mýkonos

MAP P6 ▪ Mýkonos Town ▪ 22890 28980 ▪ www. harmonyhotel.gr ▪ €€
An elegant boutique hotel, the Harmony has guest rooms that are designed to capture traditional Mýkonian styling, with features that include designer toiletries. Its à la carte restaurant uses mainly organic produce, while its guest services include beauty treatments.

Mediterranean Royal, Santoríni

MAP V2 ▪ Agía Paraskeví ▪ 22860 31167 ▪ www. mediterraneanbeach.gr ▪ €€
Skilfully designed to harmonize with the local architecture, this hotel stands in huge gardens next to a private beach. Its traditional taverna is on the beach, plus it offers guests a gym, a spa, watersports and a children's playground.

Mýconian Ambassador, Mýkonos

MAP P6 ▪ Platýs Gialós, Mýkonos Town ▪ 22890 24166 ▪ www.myconian ambassador.gr ▪ €€
Overlooking the town and its beach, this five-star hotel has elegant, themed rooms, including Arabian, Mýkonian and Asian styled rooms. Its spa, a choice of restaurants, a cocktail bar, pools and a diving centre are on site.

Cavotagoo, Mýkonos
MAP P6 ▪ Mýkonos Town
▪ 22890 20100 ▪ www.
cavotagoo.gr ▪ €€€
Dramatically built into the
rock face, this luxurious
hotel is a landmark in
the capital. Its decor is
minimalist and the rooms
are well-appointed –
some even have private
pools. The hotel has a
spa, fitness suite and
gourmet restaurant that
serves Japanese cuisine.

Madoupa Boutique, Mýkonos
MAP P6 ▪ Vrýssi village
▪ 22890 77008 ▪ www.
madoupaboutique.gr
▪ €€€
This beautiful four-star
hotel offers a luxurious
experience in the heart of
Mýkonos. Rooms come
with simple yet elegant
furnishings and are air-
conditioned, some with
private balconies.

Ostraco Suites, Mýkonos
MAP P6 ▪ Drafaki
▪ 22890 23396
▪ www.ostraco.gr ▪ €€€
A small boutique-style
retreat, the Ostraco overlooks the sea. Peaceful
sun terraces surround a
swimming pool and spa
pool, and the hotel's 21
enchanting rooms are
well-equipped to ensure
total relaxation.

Budget Stays and Camping in the Cyclades

Adonis Hotel, Náxos
MAP S4 ▪ Apóllon
▪ 22850 67060 ▪ www.
naxos-hotel-adonis.com
▪ €
This appealing blue and
white painted hotel lies
in the centre of a village,

near the beach. Its 23
rooms are attractively
furnished. The restaurant
serves breakfast, lunch
and Greek food for dinner.

Aegean Village Georgy, Páros
MAP E4 ▪ Pároikia
▪ 22840 23187 ▪ €
Standing in a pretty lemon
orchard, this little complex
of white stone cottages
looks out over the bay.
Comprising six apartments,
each with air-conditioning
and a kitchen, it is located
in the Old Town near
beaches and tavernas.

Avra Studios, Mýkonos
MAP P6 ▪ Tourlos Bay
▪ 22890 27247 ▪ €
Characterized by chic, bold
furnishings, the colour-
themed studios in this
Mýkonian country house
come with satellite
televisions, stereos and
Wi-Fi. They also have
a kitchen and an area
for dining alfresco.

Galini Pension, Íos
MAP E5 ▪ Yialos Beach
▪ 22860 91115 ▪ www.
galini-ios.com ▪ €
Situated a stone's throw
away from Yialos Beach
and the Port of Íos, this
place has standard rooms
with amenities such as air-
conditioning, TV, fridge,
hairdryer and en suite
bathrooms. Some of the
rooms afford sea views
and studios have kitchens
in the complex.

Helmos Hotel, Náxos
MAP Q4 ▪ Náxos Town
▪ 22850 22455 ▪ www.
hotelhelmos.com ▪ €
With Mediterranean
colours and extras like
fluffy towels and internet
connection, this town

centre hotel makes a
charming and good-value
place to stay. It is close
to the Áyios Geórgios beach
and nearby tavernas.

Mare-Monte Hotel, Íos
MAP E5 ▪ Seafront, Gialós
▪ 22860 91585 ▪ €
Close to a sandy beach
and fish tavernas, this
blue and white hotel offers
air-conditioned studios
that overlook the harbour,
plus a pool and a bar.

Paradise Beach Resort, Mýkonos
MAP P6 ▪ Paradise beach
▪ 22890 22852 ▪ www.
paradisemykonos.com ▪ €
This popular resort with a
campsite offers accommo-
dation in tents and simply
furnished cabins, apart-
ments and rooms. Located
on the beach, the campsite
has a restaurant, snack bar,
internet station, and lots
of watersports (see p47).

Studio Eleni, Mýkonos
MAP Q4 ▪ Agías
Paraskevís 22, Mýkonos
Town ▪ 22890 22806
▪ www.studioeleni.com ▪ €
Boasting charming
Mýkonian architecture,
this studio complex is in
the heart of town near
tavernas, bars and clubs.
The town's famous wind-
mill is minutes away.
Each studio is nicely
furnished and has air-
conditioning, internet
access and a kitchen.

Santoríni Kastelli Resort
MAP V3 ▪ Kamári
▪ 22860 31530 ▪ www.
kastelliresort.com ▪ €€
Located at the foot of
the hill of Ancient Thíra,
this charming resort

boasts five-star luxury at a reasonable price. Every room has a balcony or veranda overlooking the pools or the magnificent garden (see p47).

Luxury and Mid-Range Hotels in the Dodecanese

Konstantinos Palace, Kárpathos

MAP Y6 ▪ Pigádia ▪ 22450 23401 ▪ www. konstantinospalace.gr ▪ €
Set around its own pool and gardens close to the beach, and with amenities that include a sports complex and à la carte and buffet restaurants this is a top-end family hotel. Rooms feature lots of extras, such as music systems.

9 Muses, Pátmos

MAP F4 ▪ Sapsíla bay ▪ 22470 34079 ▪ www. 9musespatmos.com ▪ €€
An eye-catching collection of holiday cottages presented in a traditional style, the 9 Muses has a swimming pool, a bar and gardens. Gourmet dinners are delivered by a waiter to your private terrace.

Arkasa Bay Hotel, Kárpathos

MAP X6 ▪ Arkása bay, Arkása ▪ 22450 61410 ▪ www.arkasabay.com ▪ €€
Built in a style in keeping with its surroundings, this hotel offers every-thing for a pleasant stay. It has swimming pools, a children's playground, a fitness suite and restaurants that serve Greek and international dishes.

Continental Palace, Kos

MAP Y1 ▪ Georgíou Papandréou Street, Kos Town ▪ 22420 22737 ▪ www.continentalpalace. com ▪ €€
Close to the marina and ideally situated for exploring Kos Town, this modern hotel caters for families. Services include babysitting, children's pools and a playground. Most guest rooms have fine sea views and are beautifully furnished and decorated.

Esperos Village Resort, Rhodes

MAP V4 ▪ Seafront, Faliráki ▪ 22410 84100 ▪ www.esperia-hotels.gr ▪ €€
This adults-only resort offers peace and quiet away from the buzzing city of Rhodes. With plenty of restaurants to choose from, this place is perfect for a romantic dinner (see p46).

Irene Palace, Rhodes

MAP V5 ▪ Kolýmbia ▪ 22410 56224 ▪ www. irenepalace.gr ▪ €€
This family-orientated hotel provides activities that include treasure hunts and craft work-shops for children, while parents can relax at the fitness and health suite or enjoy a round of golf nearby. Its restaurant serves Rhodian fare.

Kalderimi Hotel, Astypálea

MAP F5 ▪ Livádi ▪ 22430 61120 ▪ www.kalderimi. gr ▪ €€
This luxurious hotel is housed in a traditional village cottage complex complete with arches, exposed brickwork and whitewashed walls with blue paintwork. Home-made breakfast is served on a beachside terrace.

Oceanis Hotel, Rhodes

MAP V4 ▪ Ixiá ▪ 22410 24881 ▪ www.hotel oceanis.eu ▪ €€
Overlooking the Ixiá beach, this luxurious modern hotel offers guests well-equipped rooms, a swimming pool with its own "island" bar, restaurants, an entertainment lounge and numerous sporting activities ranging from tennis to watersports.

Petra Hotel & Suites, Pátmos

MAP F4 ▪ Grikos ▪ 22470 34020 ▪ www. petrahotel-patmos.com ▪ €€
Stunning views across Grikos Bay can be enjoyed from the large and very comfortable rooms available at this lovely hotel. There is excellent dining offered and a range of on-site spa treatments.

Blue Lagoon Resort, Kos

MAP X1 ▪ Kos Town ▪ 22420 54560 ▪ www.bluelagoon resort.gr ▪ €€€
There is entertainment available for the entire family at this deluxe resort, ranging from mini soccer and a water park to the Votsalo Spa and an adults-only Italian restaurant. The rooms afford views of the sea, the pool or the gardens (see p46).

Budget Stays in the Dodecanese

Afendoulis, Kos

MAP Y1 ■ 1 Evripílou Street, Kos Town ■ 22420 25321 ■ www.afendoulis hotel.com ■ €

An inexpensive family-run hotel close to the marina and Kos Town centre, the Afendoulis has pleasingly presented air-conditioned rooms, with showers and internet access. It has a jasmine-filled garden, where breakfast is served.

Amarylis Hotel, Kárpathos

MAP Y6 ■ Pigádia ■ 22450 22375 ■ www.amarylis.gr ■ €

Well positioned close to Pigádia's shops, markets and tavernas, this pretty whitewashed hotel offers bright and airy furnished studios and apartments, each with a kitchen, private bathroom and balcony overlooking the gardens.

Asteri Hotel, Pátmos

MAP F4 ■ Skála ■ 22470 32465 ■ www.asteri patmos.gr ■ €

Close to Skála's main sights, this appealing hotel boasts traditional Pátmos architecture and is built from local stone. Rooms have sea views. There is also a breakfast terrace and gardens.

Australia Studios, Astypálea

MAP F5 ■ Péra Gialós ■ 22430 61275 ■ €

With views of the castle and located just minutes from the beach, this modern complex offers simply furnished studios with kitchens. Its garden is a delight, and there are tavernas nearby.

Galini Studios, Kálymnos

MAP F4 ■ Masoúri ■ 22430 47193 ■ www.galinistudio.gr ■ €

The island of Télendos can be seen from most of the studios in this charming pension-style guesthouse. Popular with climbers, it is close to many recognized climbing and trekking routes, as well as village tavernas.

Hotel Philoxenia, Kálymnos

MAP F4 ■ Armeos ■ 22430 59310 ■ www.philoxenia-kalymnos.com ■ €

Set in gardens of tropical palm trees close to the beach, this inexpensive, family-run hotel makes a good base for budget-conscious visitors. It has its own bar and restaurant.

Hotel Rodon, Pátmos

MAP F4 ■ Skála ■ 22470 31371 ■ €

The balconies of the attractive Hotel Rodon's well-equipped and air-conditioned guest rooms afford views of the harbour or St John's Monastery. The beach is located close by as are a number of restaurants.

Sea & Sun Studios, Rhodes

MAP U5 ■ Kiotári beach ■ 69813 77108 ■ www.kiotari.gr ■ €

Offering good value, this modern complex of air-conditioned and well-equipped holiday studios overlooks a quiet, sandy beach. Each studio has a private terrace or balcony to enjoy the view. There are tavernas nearby. Special deals are available on longer rentals.

Skala Hotel, Pátmos

MAP F4 ■ Skála ■ 22470 31343 ■ www.skalahotel.gr ■ €€

Set in a garden full of bougainvillea and tropical shrubs, this traditional Pátmos-style hotel offers guests a peaceful and relaxing place to stay. Facilities include well-presented rooms, sun decks, a conference suite and a restaurant.

Luxury and Mid-Range Hotels in the Northeast Aegean Islands

Heliotrope Hotel, Lésvos

MAP S2 ■ Vígla beach, Mytilíni ■ 22510 45857 ■ www.heliotrope.gr ■ €

Built to reflect Aegean architecture and standing in beachside gardens full of palm trees, this hotel offers rooms with sea views and amenities such as a seawater pool and hydro massage.

Kerveli Village Hotel, Sámos

MAP F4 ■ Kerveli ■ 22730 23006 ■ www.kerveli.gr ■ €

The Kerveli is a luxurious hotel located in a bay famous for its sunrises. It has six buildings surrounded by cypress and olive trees. Enjoy its pool, elegant sun terraces and its à la carte restaurant, while guest rooms are tastefully decorated.

Aeolian Village, Lésvos

MAP Q2 ■ Skála Eressoú ■ 22530 53585 ■ €€

This 84-room complex is ideal for families. Located near the beach, it offers a children's club and play area, along with pools, its own taverna and well-equipped guest rooms. Activities include Greek cooking lessons.

Filiopi Hotel, Ikaría
MAP F4 ▪ Ágios Kírykos ▪ 22750 24124 ▪ €€
A modern hotel built in traditional Aegean style only a few minutes walk from the village centre, the Filiopi comprises studios with kitchen facilities and television. Breakfast and drinks are served on its terrace.

Grecian Castle Hotel, Híos
MAP L1 ▪ Bella Vista beach, Híos Town ▪ 22710 44740 ▪ www. greciancastle.gr ▪ €€
Combining medieval Híos and modern styling, this four-star hotel is full of character. Its guest rooms are luxurious, while facilities include the à la carte Pirgos Restaurant, a cocktail bar and a pool.

Heliotrope Hotel, Argentikon Hotel, Híos
MAP L5 ▪ Kámbos ▪ 22710 33111 ▪ www. argentikon.gr ▪ €€
This complex of suites has been created within a 16th-century estate, where architectural details such as frescoes have been retained. From the restaurant and health suite to the guests' accommodation, the feel is one of total luxury.

Kalidon Panorama Hotel, Sámos
MAP F4 ▪ Kokkári ▪ 22730 92800 ▪ www. kalidon.gr ▪ €€
Guest rooms at this attractive four-star hotel,

which is built in an amphitheatre fashion around the pool, either look out over its tropical garden or towards the sea. Guests can dine on classic dishes at its terrace restaurant.

Lemnos Village Resort, Límnos
MAP E1 ▪ Platý ▪ 22540 23500 ▪ www.lemnos villagehotel.com ▪ €€
This beachside holiday resort's rooms are split-level, luxuriously furnished and well-equipped. On site are pools, sports facilities, and chic restaurants serving international cuisine (see p46).

Loriet Hotel, Lésvos
MAP S2 ▪ Variá, Mytilíni ▪ 22510 43111 ▪ www. loriet-hotel.com ▪ €€
Housed in a beautifully restored 19th-century mansion, with many of its historic features preserved, this hotel is full of character. It has elegant guest rooms, a restaurant, a courtyard pool area surrounded by pine trees and a garden. The garden has a century-old tree-lined avenue that was once used by horse-drawn carriages.

Varos Village, Límnos
MAP E1 ▪ Varos Traditional Settlement, Límnos ▪ 22540 31728 ▪ www.varosvillage.com ▪ €€
Varos is spread over six reconstructed traditional buildings. The hotel's rooms are well-equipped and it has its own taverna. The pool is the largest in the Aegean, and there is also a gym.

Budget Stays and Camping in the Northeast Aegean Islands

Agia Sion Hotel, Lésvos
MAP S2 ▪ Agiássos, near Mytilíni ▪ 22520 22242 ▪ €
In a period building amid the cobbled streets of the village centre, this small hotel offers visitors the chance to experience rural island life. Its rooms are well-equipped, and there is a garden restaurant.

Anthemis Apartments, Sámos
MAP F4 ▪ Kalami ▪ 22730 28050 ▪ www.samos-apartments.com ▪ €
This modern complex of apartments is a short walk from the beach and the village centre with its shops and tavernas. Each apartment has a telephone, a balcony and a kitchen with a fridge,

Anthemis Hotel, Ikaría
MAP F4 ▪ Thermá ▪ 22750 23156 ▪ www. anthemishotelikaria.com ▪ €
This two-star hotel offers super guest rooms that come complete with air-conditioning, a dining terrace and a fridge. Room service is offered. The hotel is close to the beach and the village centre.

Chíos Camping, Híos
MAP K5 ▪ Ágios Isidóros-Sykiáda ▪ 22710 74111 ▪ €
In a waterside location, facilities here include a restaurant, shops and a launderette. This is one of the more organized campsites on Híos, which has electrical hook-ups and children's play areas.

Dionyssos Campsite, Lésvos

MAP R3 ▪ Vaterá ▪ 22520 61151 ▪ €

This pleasant campsite offers pitches for tents and caravans near the popular beach at Vaterá. On-site facilities include electrical hook-up, water coolers and showers. There is a designated building for cooking and a laundry room for guests use. Nearby there are tavernas, bars and shops.

Hotel Cohyli, Sámos

MAP F4 ▪ Seafront, Iréon ▪ 22730 95282 ▪ www. hotel-cohyli.com ▪ €

Enjoy breakfast on the terrace of this small family-run hotel while listening to the sound of the lapping waves. Attractively decorated guest rooms, the superb Garden Restaurant and Greek hospitality all contribute to making this a memorable place.

Lymberis Apartments, Límnos

MAP E1 ▪ Mýrina ▪ 22540 23352 ▪ www.despina lymperihotel.com ▪ €

A great place to stay for anyone on a budget, these air-conditioned apartments are close to the beach and harbour. They come complete with a kitchen and a minibar, and there is a breakfast terrace on site.

Plaka Studios, Híos

MAP L5 ▪ Karfás ▪ 22710 32955 ▪ www. plakastudios.gr ▪ €

Offering great value, Plaka Studios are housed in a new, stone-built complex that stands in nice gardens overlooking the sea. The 10 studios have air-conditioning, modern well-equipped kitchens, a fridge and satellite television.

Zorbas Apartments, Híos

MAP K4 ▪ Volissós ▪ 22740 21436 ▪ www. chioszorbas.gr ▪ €

Run by Yiannis Zorbas and his family, this long-established complex of apartments is a gem. The studios are beautifully clean and stylish, and have a TV, internet access and air-conditioning. All are also fully-equipped with kitchens and bathrooms with showers. The meals are local dishes and guests sit down to eat with the family.

Luxury and Mid-Range Hotels in the Sporádes and Évvia

Adrina Beach Hotel, Skópelos

MAP M1 ▪ Pánormos ▪ 24240 23373 ▪ www.adrina.gr ▪ €€

A four-star hotel, the Adrina is close to the beach. Features include a spa pool that overlooks the sea and an elegant terrace taverna serving Greek dishes.

Aegean Suites Hotel, Skiáthos

MAP M1 ▪ Megáli Ámmos ▪ 24270 24066 ▪ www. aegeansuites.gr ▪ €€

A five-star boutique hotel, the Aegean oozes understated luxury. From lavish fabrics to individual music centres, the guest rooms are outstanding. Amenities include a gourmet restaurant, a gym and a holistic beauty centre. The hotel is exclusively for adults and teenagers.

Almira Mare Hotel, Évvia

MAP M3 ▪ Ágios Minás, Halkída ▪ 22210 97100 ▪ www.almiramare.gr ▪ €€

This modern four-star hotel has a children's playground and pool, an adult's pool, sports amenities and restaurants. All rooms have internet and air-conditioning.

Apollon Suites Hotel, Évvia

MAP D3 ▪ Kárystos ▪ 22240 22045 ▪ www. apollonsuiteshotel.com ▪ €€

This four-star complex of 36 sea-view suites is close to the beach and the harbour, and not far from the centre of town either. The guest rooms are well-equipped. There is an à la carte restaurant, sports amenities and a children's playground.

Atrium Hotel, Skiáthos

MAP M1 ▪ Agía Paraskeví, Plataniás ▪ 24270 49345 ▪ www.atriumhotel.gr ▪ €€

This 75-room luxury hotel is built of natural stone in amphitheatre style on a hillside overlooking Agía Paraskeví beach. Its many amenities include chic, well-equipped guest rooms, a taverna serving homemade local dishes, swimming pools, a spa and ball sports.

Esperides Hotel, Skiáthos

MAP M1 ▪ Skiáthos Town ▪ 24270 22245 ▪ www. esperidesbeach.gr ▪ €€

A modern hotel with 180 rooms, the Esperides is a beachside complex within easy reach of the town centre. It is family friendly, with special play areas and menus for small appetites. Facilities include a pool, a restaurant and internet access.

Hotel Nefeli, Skýros

MAP P2 ▪ Skýros Town ▪ 22220 91964 ▪ www. skyros-nefeli.gr ▪ €€
A sense of luxury envelops you as you step into this eco hotel built around a sea-water swimming pool and elegant sun terraces. It has an à la carte breakfast bar and a taverna.

Negroponte Resort, Évvia

MAP D3 ▪ Erétria ▪ 22290 61935 ▪ www. negroponteresort.gr ▪ €€
The friendly service, spacious rooms and proximity to the beach make this hotel a perfect option for families. The exceptional cuisine served in the restaurant and indulgent treatments available at the spa make this resort popular with travellers (see p46).

Pleoussa Studios, Skópelos

MAP M1 ▪ Skópelos Town ▪ 24240 23141 ▪ www. pleoussa-skopelos.gr ▪ €€
The Pleoussa Studios is a collection of 10 luxury studios built with paved areas and arches to reflect the architecture of Skópelos. Each suite has pleasing decor, air-conditioning and a kitchen. Internet access is provided. Each unit has panoramic views of the bay.

Skíathos Palace Hotel, Skíathos

MAP M1 ▪ Koukounariés ▪ 24270 49700 ▪ www. skiathos-palace.gr ▪ €€
With views of the famous Koukounariés beach, this five-star hotel is one of the most luxurious on the island. Its 258 rooms are very tasteful and the rooftop restaurant is known for its creative cuisine. It also has tennis and children's play areas.

Budget Stays and Camping in the Sporádes and Évvia

Aegeon Hotel, Skópelos

MAP M1 ▪ Skópelos Town ▪ 24240 22619 ▪ €
This 15-room, two-star modern hotel, built in traditional style, is covered in bougainvillea and stands on a hillside overlooking the bay. The rooms are simple, well-equipped and have a balcony, television and air-conditioning. There is a children's playground too.

Hotel Pothos, Skíathos

MAP M1 ▪ Skíathos Town ▪ 24270 22694 ▪ www. pothos-skiathos.com ▪ €
A small, attractive hotel in the heart of Skíathos Town, the Pothos stands in mature gardens of palms and red hibiscus. Rooms are pleasingly furnished and have air-conditioning, plus balconies that look out over the gardens.

Hotel Regina, Skópelos

MAP M1 ▪ Skópelos Town ▪ 24240 22138 ▪ €
Although the amenities of this two-star hotel are basic, it has a great side-street location looking over the bay. There is a charming breakfast room, while its 11 modern bedrooms have air-conditioning and fridges.

Karakatsanis Nikos Rooms, Alónissos

MAP N1 ▪ Vótsi ▪ 24240 66188 ▪ €
This collection of rooms makes a good, inexpensive base from which to explore the island. Each has a fridge, a private bathroom and a balcony. There is a community kitchen and a courtyard dining area. All rooms have glorious sea views.

Karystion Hotel, Évvia

MAP E3 ▪ Kárystos ▪ 22240 22391 ▪ www. karystion.gr ▪ €
Tastefully decorated, this two-star hotel represents good value. It stands on a peninsula with great views of the bay. You can dine inside or on its "wrap-around" terrace and relax in the coffee bar. Rooms have air-conditioning.

Koukounariés Camping, Skíathos

MAP M1 ▪ Koukounariés beach ▪ 24270 49250 ▪ €
This campsite is found just off the beach and has a restaurant, supermarket and children's play areas. There are designated plots under the canopy of the camp's mature trees. A launderette is on site, as are showers.

Pandora Studios, Skíathos

MAP M1 ▪ Koliós ▪ 24270 49272 ▪ www.skiathos pandora.gr ▪ €
The beautifully presented apartments and studios here have sea views and a kitchen, satellite television, internet access and air-conditioning. They are located in gardens with palm trees near the village centre and beach.

For a key to hotel price categories see p166

Pension Maria, Alónissos

MAP N1 ■ Patitíri ■ 24240
65348 ■ www.alonnisos-
maria.com ■ €
Surrounded by pine trees
and close to the bustling
harbourside, this collec-
tion of studios offers great
views and good amenities.
Each studio has a kitchen
with a fridge, a private
bathroom and a balcony.
Tavernas, bars and
restaurants are nearby.

Rovies Camping, Évvia

MAP D3 ■ Rovies, near
Límni ■ 22270 71120 ■ €
This campsite opened
in 1985 and has a loyal
following. It is located
amid gardens and trees
at the foothills of Mount
Telethrion and right next
to the beach. Amenities
include plots for tents and
caravans, a supermarket,
a launderette, a café and
a children's playground.

Valledi Village Hotel, Évvia

MAP N3 ■ Kými ■ 22220
29150 ■ www.valledi
village.gr ■ €
A swimming pool with
attractive sun terraces
and a breakfast lounge
are just two features of
this modern two-star
apartment-style hotel
overlooking the bay.
Rooms have kitchens and
are stylishly decorated.

Luxury and Mid-Range Hotels in the Argo-Saronic Islands

Hotel Brown, Égina

MAP K1 ■ Égina Town
■ 22970 22271 ■ www.
hotelbrown.gr ■ €
The 19th-century sea
sponge factory that was

transformed into Hotel
Brown is an Égina land-
mark. This elegant hotel
comprises 28 rooms,
with great views of the
bay, and a collection of
bungalows that are dotted
around the gardens.

Sto Roloi Hotel, Póros

MAP K3 ■ Kostelénou
34-36 ■ 22980 25808
■ www.storoloi-poros.gr
■ €
Housed in a beautifully
renovated 19th-century
Neo-Classical building,
Sto Roloi, meaning "At
the Clock Tower", is a
collection of luxurious
apartments and a suite.
Elegantly furnished with
antiques, the rooms have
features such as massage
showers. The hotel is
close to the harbour.

Bratsera, Hydra

MAP D4 ■ Hóra, Hydra
Town ■ 22980 53971
■ www.bratserahotel.com
■ €€
Named after a local ship
used by sponge divers
and housed in a charming
restored 19th-century
sponge factory, this hotel
is luxurious. The specta-
cular lobby leads through
gardens to the pool, restaurant and 26 rooms.

Economou Mansion, Spétses

MAP D4 ■ Harbourside,
Spétses Town ■ 22980
73400 ■ www.economou
spetses.gr ■ €€
This 19th-century former
naval master's mansion
features period archi-
tecture and antiques.
The luxurious studios
and suites all have air-
conditioning and Wi-Fi.
Some have private
terraces with sea views.

Hotel Angelica, Hydra

MAP D4 ■ Andréa Miaoúli
43, Hydra Town ■ 22980
53202 ■ www.angelica.gr
■ €€
This boutique hotel is
housed in an old mansion.
Its original architecture
of stone walls, wooden
beams and arches gives
its well-decorated rooms
much character. Among
its features are a sauna, a
pool and a business suite.

Leto Hotel, Hydra

MAP D4 ■ Hydra Town
■ 22980 53385
■ www.letohydra.gr ■ €€
A total of 21 rooms,
including four that are
interconnected and one
with wheelchair access,
are offered at this family-
orientated hotel. A lounge,
a gym and a garden pro-
vide places to unwind.

Mistral Hotel, Hydra

MAP D4 ■ Harbour, Hydra
Town ■ 22980 52509
■ www.hotelmistral.gr ■ €€
A luxury hotel in a country
mansion-style building,
the Mistral offers 18 individually designed rooms
with quality bathrooms
and features like satellite
television. Breakfast is
served out in the garden
and includes delicious
homemade bread.

Orloff Resort, Spétses

MAP D4 ■ Old Harbour,
Spétses Town ■ 22980
75444 ■ www.orloff
resort.com ■ €€
Chic, minimalist decor,
handmade furniture,
crisp linens and marble
bathrooms are some of
the features of the suites
and maisonettes at this
resort. It uses natural
materials to reflect traditional architecture.

Vasilis Bungalows, Kýthira

MAP C5 ▪ Kapsáli
▪ 27360 31125 ▪ www.
kithira.biz ▪ €€

A collection of 12 bunga-
lows built among mature
olive trees, this complex
is ideal for exploring
Hóra. The well-furnished
bungalows have air-
conditioning and a private
patio. Guests can enjoy
breakfast on the terrace.

Zoe's Club Hotel, Spétses

MAP D4 ▪ Harbourside,
Spétses Town ▪ 22980
74447 ▪ www.zoesclub.gr
▪ €€

Close to the beach and
the port of Dápia, this
charming complex of
apartments has been
built around a pool and a
sun terrace. The luxury
guests' accommodation
has every amenity, from
bathrobes to DVD players
and internet access.

Budget Stays in the Argo-Saronic Islands

7 Islands Hotel, Spétses

MAP D4 ▪ Ag. Marina,
Spétses Town ▪ 22980
73059 ▪ www.7islands-
spetses.com ▪ €

On a hillside overlooking
the harbour, this Anglo–
Greek-owned hotel is
10–15 minutes' walk from
the main town. It provides
a peaceful setting for a
stay in Spétses.

Delfini Hotel, Hydra

MAP D4 ▪ Hydra Town
▪ 22980 52082 ▪ www.
delfinihotel.gr ▪ €

This small hotel is set
in a cluster of traditional
Hydran seamen's cottages
overlooking the bay. Fully

renovated, its rooms
are spacious and well
equipped, many boasting
sea views. The terrace
is a great place to watch
the bustle of the harbour.

Hotel Liberty 2, Égina

MAP L1 ▪ Agia Marína
▪ 22970 32105 ▪ www.
hotelliberty2.gr ▪ €

Covered in bougainvillea
and boasting traditional
architecture, this pleasing
hotel is just minutes from
the beach. Its inviting
guest rooms look out over
the bay, while facilities
include a breakfast
terrace, a gift shop and a
satellite television lounge.

Hotel Plaza, Égina

MAP K1 ▪ Égina Town
▪ 22970 25600 ▪ www.
aeginaplazahotel.com ▪ €

This small, welcoming
hotel looks out over
the harbour. Its guest
studios each have a fully
equipped kitchen with
fridge, enabling self-
catering. It also has
air-conditioning and a
television in each room,
while a whole raft of
tavernas, bars and
shops are within walking
distance. There is also
a laundry room on site.

Hydroussa Hotel, Hydra

MAP D4 ▪ Platía Vótsi,
Hydra Town ▪ 22980
52400 ▪ www.hydroussa-
hydra.gr ▪ €

The Hydroussa is housed
in a traditional mansion-
style property within
walking distance of the
harbour. Guests can
enjoy air-conditioned
rooms and dine in the
breakfast hall or relax
in the drawing room or
the picturesque gardens.

Kamares Apartments, Kýthira

MAP C5 ▪ Aroniádika
▪ 27360 33420 ▪ www.
apartmentskamares.com
▪ €

This complex of 10
apartments is located in
an 18th- and 19th-century
rural house, whose stone
arches give it plenty of
character. There's air-
conditioning, parking and
private kitchens. Great
views of the countryside.

Margarita Hotel, Kýthira

MAP C5 ▪ Hóra ▪ 27360
31711 ▪ www.hotel-
margarita.com ▪ €

The elevated location of
this hotel ensures it has
panoramic views from
most rooms. Housed in
a 19th-century mansion,
it offers good value with
12 air-conditioned rooms
and a terrace where
breakfast is prepared
by the French owners.

Pension Electra, Égina

MAP K1 ▪ Égina Town
▪ 22970 26715 ▪ www.
aegina-electra.gr ▪ €

Despite its location in
the centre of town, this
family-run pension is
quiet and relaxing as well
as clean and comfortable.
First-floor studios have
sea views, and each has
a terrace or a patio area.

Pension Erofili, Hydra

MAP D4 ▪ Hydra Town
▪ 22980 54049 ▪ www.
pensionerofili.gr ▪ €

A stone cottage minutes
from the harbourside, this
pension is a convenient
base for exploring Hydra.
It offers 11 tastefully decorated and air-conditioned
rooms and one suite. The
paved courtyard is sur-
rounded by bougainvillea.

For a key to hotel price categories see p166

Villa Rodanthos, Égina

MAP K2 ■ Pérdika
■ 22970 61400 ■ www.
villarodanthos.com ■ €
A pretty Neo-Classical building with 10 self-catering studios, the Villa Rodanthos is just minutes from the harbour, which is known for its fish tavernas. It has a breakfast lounge, a bar and a roof garden with a fine view out to the sea. Each studio has a kitchenette and a balcony.

Luxury and Mid-Range Hotels in Crete

Anissa Beach Hotel

MAP E6 ■ Seafront, Anissáras, Chersónissos
■ 28970 23264 ■ www.
anissabeach.com ■ €
Enjoy watersports, playing tennis, swimming and eating Mediterranean food in the La Pergola restaurant at this hotel overlooking the sandy beach of Anissáras. Its rooms are chic and well-equipped, most having sea views.

AKS Annabelle Beach Resort

MAP E6 ■ Seafront, Chersónissos ■ 28970 23561 ■ www.annabelle beachresort.gr ■ €€
This five-star village hotel lies next to a Blue Flag beach and offers pools, watersports and a fitness centre with a sauna and treatment room. The 262 bungalow rooms are luxurious (see p47).

Albatros Hotel

MAP E6 ■ Seafront, Chersónissos ■ 28970 22144 ■ www.albatros.gr ■ €€
An attractive whitewashed hotel standing in lush gardens of palm trees and tropical shrubs, the Albatros offers a choice of dining areas, including its super à la carte Daphne restaurant. On site are a children's club and pool, and a fitness suite.

Angela Suites Boutique Hotel

MAP E6 ■ Sísi village
■ 28410 71121 ■ www.
angelasuites.com ■ €€
Contemporary and stylish, this collection of suites lies close to Sísi beach and the village. Individually designed rooms in vivid colours are chic and well-equipped, while facilities include freshwater pools, a fitness and alternative therapy suite, a cocktail lounge and an à la carte poolside restaurant.

Fodele Beach and Water Park

MAP E6 ■ Irákleio
■ 28105 22000 ■ www.
fodelebeach.gr ■ €€
This lively resort offers bungalow guest rooms that are built in a style reminiscent of Venetian architecture. In a beautiful setting with a private beach, this place has every amenity desired, from restaurants serving European cuisine to bars, a children's club, sports, a gym and an on-site waterpark.

Galaxy Hotel

MAP E6 ■ Leofóros Dimokratías 75, Irákleio ■ 28102 38812 ■ www.
galaxy-hotel.com ■ €€
One of the finest hotels on the island, the Galaxy is the last word in luxury. Amenities at this 127-room hotel include gourmet restaurants, with indoor and alfresco options, and a wellness centre with a hammam and a gym that is complimentary for guests.

Hersonissos Palace

MAP E6 ■ Harbourside, Chersónissos ■ 28970 23603 ■ www.hersotels.gr ■ €€
This hotel, with a mix of Greek Neo-Classical and contemporary styling, has 150 rooms, four suites and tennis courts. Mediterranean cuisine is served in its restaurant and poolside buffet.

Istron Bay Hotel

MAP F6 ■ Harbourside, Ístro, Ágios Nikólaos
■ 28410 61303 ■ www.
istronbay.gr ■ €€
Overlooking a bay, this hotel has a welcoming feel. It celebrates Cretan traditions with music evenings and Greek à la carte cuisine, while offering amenities such as watersports, tennis and aerobics.

Candia Maris Resort & Spa, Crete

MAP E6 ■ Andréa Papandréou Street 72, Irákleio ■ 28103 77000 ■ www.candiamaris.gr ■ €€€
This charming resort and spa with tasteful rooms is housed in six separate buildings set in lush gardens. Smoking and non-smoking rooms are available, but guests can also choose from deluxe rooms or suites that offer a luxury pool and terrace (see p47).

Elounda Bay Palace

MAP F6 ■ Seafront, Eloúnda ■ 28410 67000 ■ www.
eloundabay.gr ■ €€€
Surrounded by large gardens and with an

infinity pool overlooking the sea, this hotel is a haven of peace. There is a water's edge restaurant, a private gym and a spa.

Budget Stays and Camping in Crete

Adelais Hotel

MAP D6 ▪ Tavronítis, Chaniá ▪ 28240 22929 ▪ www.adelais.gr ▪ €
Constructed in the style of the local stone homes, this super hotel has rooms decorated in a bright Mediterranean theme, sun terraces, pools and a restaurant serving refined cuisine. There is a children's playground and pool.

Camping Sísi

MAP E6 ▪ Sísi, Lasíthi ▪ 28410 71247 ▪ www. sisicamping.gr ▪ €
The excellent facilities at this attractively located campsite include a swimming pool, Wi-Fi, a laundry room and a well-equipped kitchen. On the northeast coast of Crete, it's just a 15-minute walk to the pretty village of Sisi, and only a 5-minute walk to the beach.

Hotel Castro Amoudára

MAP E6 ▪ A Papandréou Street 301, Amoudára ▪ 28108 22770 ▪ www. castro-hotel.com ▪ €
This attractive hotel has 53 guest rooms, most of which have a balcony looking out over the pool and gardens. Maxims, its own traditional-style taverna, serves Cretan cuisine and wine, while a pool bar serves cocktails.

Hotel Kissamos

MAP D6 ▪ Iroón Polytechníou 172, Kíssamos ▪ 28220 22086 ▪ www.hotelkissamos.gr ▪ €
Comprising 24 rooms with private bathrooms, air-conditioning and great views of Kíssamos bay, this hotel offers guests a home-from-home feel. There are some palm-tree-shaded gardens and a swimming pool available.

Lefkoniko Seaside Hotel

MAP E6 ▪ Sofoklí Venizélou 76 and Eleftherías, Réthymno ▪ 28310 55326 ▪ €
This complex of air-conditioned studios and apartments has sea views and is central for local amenities. Buffet and à la carte restaurants, bars and swimming pools are available at its nearby sister hotel, the Lefkoniko Beach Hotel.

Mirtilos Studios-Apartments

MAP D6 ▪ Tzanakákis Square, Kíssamos ▪ 28220 23079 ▪ www.mirtilos. com ▪ €
Centrally located minutes from the beautiful sandy beach of Telonío, with views of Kíssamos bay, this complex of well-equipped studios and apartments is a good holiday base. It has nice gardens with sun terraces, a dining area and two large pools.

Mithimna Camping

MAP D6 ▪ Drapanías, Kíssamos ▪ 28220 31444 ▪ www.camping mithimna.com ▪ €
This large campsite is well organized

and features a good restaurant serving tasty Cretan dishes. Other facilities include a laundry, hot and cold water and electrical hook-up. There is also a mini-supermarket on site. Caravan and tent owners are both welcome.

Réthymno Youth Hostel

MAP E6 ▪ Tobázi Street 41, Réthymno ▪ 28310 22848 ▪ www.yh rethymno.com ▪ €
Housed in an attractive Venetian stone building standing amid gardens in the heart of Réthymno, life at this youth hostel centres around its pretty patio. Dormitory rooms are basic but clean, and internet and light meals are available.

Youth Hostel Plakias

MAP E6 ▪ Plakiás ▪ 28320 32118 ▪ www.yhplakias. com ▪ €
Nestled in an olive grove with walking trails and beaches nearby, this great chilled-out hostel provides nicely furnished shared bungalows, inexpensive meals, drinks and Wi-Fi access. It also hosts fun barbecues and a series of party nights.

Hotel Caretta Beach

MAP D6 ▪ Geráni Kydonías, Chaniá ▪ 28210 61700 ▪ www.caretta-beach.gr ▪ €€
Offering good value for money, these apartments are beautifully furnished and come with room service and internet access. The hotel is surrounded by lovely olive groves, and has a pool, a restaurant and a beach.

For a key to hotel price categories see p166

Index

Acknowledgments

Author
Carole French is an award-winning BBC-trained journalist, based in Cyprus and the UK. Her work has appeared in publications including *ABTA Magazine*, *Homes Overseas* and the *Daily Mail*. She has worked on travel guides for Michelin, Time Out and Thomas Cook, and provided expert consultation on the Greek Islands for television.

Publishing Director Georgina Dee

Publisher Vivien Antwi

Design Director Phil Ormerod

Editorial Sophie Adam, Ankita Awasthi Tröger, Rebecca Flynn, Rachel Fox, Alison McGill, Sally Schafer, Jackie Staddon

Cover Design Richard Czapnik

Design Hansa Babra, Tessa Bindloss, Vinita Venugopal

Commissioned Photography Max Alexander, Paul Harris, Rough Guides / Michelle Grant, Helena Smith Tony Souter

Picture Research Susie Peachey, Ellen Root, Lucy Sienkowska

Cartography Stuart James, Suresh Kumar, Casper Morris, Reetu Pandey

DTP Jason Little

Production Stephanie McConnell

Factchecker Mariana Evmolpidou

Proofreader Leena Lane

Indexer Hilary Bird

Picture Credits

The publisher would like to thank the following for their kind permission to reproduce their photographs:
Key: a-above; b-below/bottom; c-centre; f-far; l-left; r-right; t-top

123RF.com: Lucian Bolca 114b; freeartist 154t; Dariya Maksimova 4cl.

Alamy Stock Photo: Acorn 1 80c; Prisma Archivo 33tr, 146bl; Art of Travel 57tr; George Atsametakis 149tl; Erin Babnik 32crb; Gary Blake 140b; Sergey Borisov 146t; David Crossland 70cla; Ian Dagnall 80bl; Adam Eastland 64br; Eyre 98tc; Malcolm Fairman 60bl; funkyfood London - Paul Williams 22–3, 86cla; Olga Gajewska 39br; Granger Historical Picture Archive 38b; Hackenberg-Photo-Cologne 94tc; Terry Harris 61cla, 71tl; Peter Horree 33bc; Constantinos Iliopoulos 4crb, 64t, 150cr; imageimage 17tl; Images&Stories 21tl; IML Image Group /George Detsis 96b; Pawel Kazmierczak 24bl; LOOK Die Bildagentur der Fotografen GmbH /Ingolf Pompe 65crb; Hercules Milas 20cla, 25tl, 28tl, 30cla, 34–5, 79br, 96tl, 106tl, 116br, 132br; nagglestock. com: 19tl; Nikos Pavlakis 145tl, 150tl; Pictures Colour library 58t; Rob Rayworth 129br; REDA &CO srl /Riccardo Lombardo 14b; Simon Reddy 129c; Rolf Richardson 134tl; Chris Rout 120tl; Peter Schickert 142br; Charles Stirling (Diving) 49tr, (Travel) 46cl; Petr Svarc 31cb; Ian Thraves 70b; Travel Pictures /Pictures Colour Library 78tl; TravelCollection /Arthur F. Selbach 45tr; Urbanmyth 132tl; Aristidis Vafeiadakis 11tr; Vito Arcomano Photography 50clb; Terence Waeland 141cl; Ian West 67tr, 136tl; YAY Media AS /Arsty 29clb.

AWL Images: Hemis 92b.

Basilico: 80cr.

The Captain's Table: 133cl.

CRETAquarium: 59tr.

Crete Golf Club: 56clb.

Dorling Kindersley: Nea Moni /Helena Smith 11tl; Heraklion Archaeological Museum /Tony Souter 42ca; Monastery of St John, Patmos / Tony Souter 10crb, 16br.

Dreamstime.com: Adisa 52t; Rostislav Ageev 102c; Airphoto 135t, 138br; Amlyd 6l; Anilah 49clb; Arenaphotouk 26c; Arsty 102t; Asteri77 7tr; Mila Atkovska 18cla, 41br, 48br, 52bl, 54–5, 68t, 76cr, 76–7, 138cl; Axpitel 104clb; Helena Bilková 26bl; Junior Braz 10cl; Dan Breckwoldt 10cla; Bstefanov 94clb; Volodymyr Byrdyak 111cra; Ccat82 15br; Charlieaja 108c; Chasdesign1983 60t; Anton Chygarev 11cr; Coplandj 110tl; Paul Cowan 11clb, 32–3, 62tl, 87cb; Mario Cupkovic 139t,;Tomasz Czajkowski 137cla; Dbyjuhfl 26–7; Debu55y 15tl, 116cl; Igor Dutina 62cb; Dziewul 27cr, 56tr, 103bl, 106br; Earthonware 147clb; Denis Fefilov 69tr; Inna Felker 66tr; Flavijus 67cl; David Elliott 59bl; Fotokon 40t; Foton64 79cl; David Fowler 135br; Freesurf69 1, 4t, 90tr, 90–91, 98bl, 118b, 136b; Gordon Bell 86b; Gyuszko 51b; H368K742 11ca; Tal Hayoun 143tl; Stoyan Haytov 50tr; Hlphoto 62br; Imagin.gr Photography 57b, 148b; Irishka777 69tl; Liliya Kandrashevich 63tr; Tzogia Kappatou 61br; Panagiotis Karapanagiotis 2tl, 8–9, 16–17, 18–19, 82tl, 126–7; Karelgallas 54cla; Aleksandrs Kosarevs 53clb; Patryk Kosmider 105b; Lejoch 30br; Lornet 66b; Lucianbolca 84br; Dariya Maksimova 34clb; Tomas Marek 48t; Ivan Mateev 63cl; Milosk50 12cla, 12–13; Mineria6 40bc; Mkos83 55tr; Martin Molcan 44tl; Nanookofthenorth 35bl; Nomadbeg 29crb; Olgacov 4b; Lefteris Papaulakis 35tl; Olena Pavlovich 13br; Photocreation 66c; Photostella 11br, 144tl; Rostislavv 104tr; Alex Saluk 113tl;

Scorp37 30–31; Selitbul 44–5; Serrnovik 14–15; Slasta20 12br; Smallredgirl 27tl; Smoxx78 75tc; Spectrumoflight 88tl; Andrei Stancu 10b; Stockbksts 7cl, 24–5, 124tl; Svetap 43tl; Topolov 53tr; Totophotos 55br; Travelbook 68br; Cristina Trif 74ca; Manolis Tsantakis 109tr; wabeno 82bc; Witr 89tr; Xiaoma 4cla, 29b, 31tr.

FLPA: Imagebroker 84tl.

Getty Images: Print Collector 41cl; DEA /G. Dagli Orti 39cla; Lonely Planet Images /Izzet Keribar 25cr, 45cla, 125tl, 130cla; Minden Pictures /Konrad Wothe 126bc; George Papapostolou 120clb.

iStockphoto.com: Angelika 4cra; Arsty 101tr; Bahan19 97cr; Banet12 100cl; DavidCallan 122–3; Gatsi 145br; Guenter Guni 2tr, 36–7; KarelGallas 4clb; Starcevic 3tr, 156–7b; Vladimir_Timofeev 112b; wwalakte 3tl, 72–3.

Koukoumavlos Restaurant: 99clb.

Lofaki Restaurant: 121cr.

Lorraine's Magic Hill: 87tr.

Nontas Fish Restaurant: 155clb.

Piraeus Bank Group Cultural Foundation/ Museum of Marble Crafts: Nikos Daniilidis 43br.

Rex Shutterstock: Working Title / Canal + / Universal /Peter Mountain 71br.

Robert Harding Picture Library: Ashley Cooper 125br; Neil Farrin 42b; Philippe Michel 17crb; Carlo Morucchio 126clb; Jose Fuste Raga 128t; Ellen Rooney 51cra.

Spilia Seaside Restaurant & Cocktail Bar: 95cr.

SuperStock: age fotostock 19cra; Funkystock /age fotostock 20–21, 20br, 21crb.

Taverna Karbouris: 83cr.

Yacht Club Panagakis: 151crb.

Zakanthi Restaurant: 85cr.

Cover
Front and spine: **4Corners:** Colin Dixon; back: **123RF.com:** ccat82.

Pull Out Map Cover
4Corners: Colin Dixon.
All other images © Dorling Kindersley
For further information see:
www.dkimages.com

Penguin
Random
House

Printed and bound in China

First published in Great Britain in 2011
by Dorling Kindersley Limited
80 Strand, London WC2R 0RL

Copyright 2011, 2018 © Dorling
Kindersley Limited

A Penguin Random House Company

18 19 20 21 10 9 8 7 6 5 4 3 2 1

**Reprinted with revisions
2013, 2015, 2018**

A CIP catalogue record is available
from the British Library.

ISBN 978 0 2413 0917 9

MIX
Paper from
responsible sources
FSC™ C018179
www.fsc.org

SPECIAL EDITIONS OF DK TRAVEL GUIDES

DK Travel Guides can be purchased in bulk quantities at discounted prices for use in promotions or as premiums. We are also able to offer special editions and personalized jackets, corporate imprints, and excerpts from all of our books, tailored specifically to meet your own needs.

To find out more, please contact:

in the US
specialsales@dk.com

in the UK
travelguides@uk.dk.com

in Canada
specialmarkets@dk.com

in Australia
**penguincorporatesales@
penguinrandomhouse.com.au**

As a guide to abbreviations in visitor information blocks: **Adm** = admission charge; **DA** = disabled access; **D** = dinner; **L** = lunch.

Selected Map Index